COAL PYROLYSIS

COAL SCIENCE AND TECHNOLOGY

Series Editor:

Larry L. Anderson
Department of Mining, Metallurgical and Fuels Engineering, University of
Utah, Salt Lake City, UT 84112, U.S.A.

COAL SCIENCE AND TECHNOLOGY 4

COAL PYROLYSIS

G.R. GAVALAS

Division of Chemistry and Chemical Engineering,
California Institute of Technology,
Pasadena, California, U.S.A.

ELSEVIER SCIENTIFIC PUBLISHING COMPANY
Amsterdam – Oxford – New York   1982

ELSEVIER SCIENTIFIC PUBLISHING COMPANY
Molenwerf 1
P.O. Box 211, 1000 AE Amsterdam, The Netherlands

Distributors for the United States and Canada:

ELSEVIER SCIENCE PUBLISHING COMPANY INC.
52, Vanderbilt Avenue
New York, N.Y. 10017

Library of Congress Cataloging in Publication Data

Gavalas, George R.
    Coal pyrolysis.

    (Coal science and technology ; 4)
    Bibliography: p.
    Includes index.
    1. Coal. 2. Pyrolysis. I. Title. II. Series.
TP325.G38 1982   662.6'2   82-11374
ISBN 0-444-42107-6

ISBN 0-444-42107-6 (Vol. 4)
ISBN 0-444-41970-5 (Series)

Printed in The Netherlands

PREFACE

Prompted by the need of non-petroleum-based fuels, coal research has reemerged to center stage after a lengthy dormant period. Pyrolysis research, in particular, has gained considerable momentum because of its close connection to combustion, hydropyrolysis and liquefaction. Spectroscopic and other instrumental techniques are currently producing prodigious information about coal structure and pyrolysis mechanisms, while modeling efforts are breaking new ground in sorting out chemical and physical phenomena to provide a fundamental although simplified description.

The continuing generation of experimental data will lead to revisions, in some cases drastic, of current structural and kinetic precepts. Yet, the postulates and assumptions of current work provide a meaningful starting point in elaborating theoretical descriptions of greater validity and applicability.

This monograph was written to organize recent results of pyrolysis research. Experimental and theoretical aspects, given approximately equal weight, are discussed in the light of basic chemical and physical mechanisms. With this orientation the monograph should be useful to chemists, engineers and graduate students with interests in coal research.

I would like to express my appreciation to the copyright holders for permission to reproduce various figures: to the American Institute of Chemical Engineers for Figs. 4.21, 5.7, 5.14; to the Combustion Institute for Figs. 4.3, 4.14, 4.15; to IPC Science and Technology Press, Ltd. for Figs. 4.16-4.18, 4.20, 5.1-5.6, 5.8, 5.9, 6.1, 6.6, 7.10-7.13; to Dr. W. R. Ladner for Figs. 7.6-7.9; to Dr. P. R. Solomon for Fig. 4.1; to Mr. M. Steinberg for Figs. 7.14-7.16 and to Dr. E. M. Suuberg for Figs. 5.10, 5.11, 6.2-6.5, 7.1-7.4.

I would also like to thank Lenore Kerner and Majda Andlovec for typing the manuscript and Heather Marr for drawing the figures.

Pasadena, California
April, 1982   G. R. GAVALAS

TABLE OF CONTENTS

TABLE OF CONTENTS (CONTINUED)

Chapter 1

INTRODUCTION

During the last decade, developments of both experimental and theoretical nature
have advanced our knowledge of coal pyrolysis to a degree that justifies a systematic exposition on the subject. Noteworthy among experimental advances and refine-
ments have been the design of equipment for rapid but controlled heating and the
analysis and characterization of pyrolysis products by chromatographic and spectroscopic measurements. Modeling advances worth mentioning include the estimation
of rate parameters by group additivity techniques, improved capabilities for
numerical handling of complex reaction systems and improved understanding of the
coupling between mass transfer and chemical reactions.

Despite the aforementioned advances, the bulk of available information is of
experimental and descriptive nature while the understanding of fundamental mechanisms remains largely qualitative. In surveying and summarizing experimental information, we have been selective rather than exhaustive, emphasizing studies
involving broad and controlled variation of experimental conditions. The discussion of experimental results and kinetic models has concentrated on fundamental
issues such as reaction mechanisms and interaction between reactions and mass
transfer. This choice of subject matter clearly reflects the personal interests
of the author.

Early pyrolysis research was concentrated on coal carbonization, i.e. slow heating of dense coal samples, conditions pertaining to coke making. By contrast,
recent work has emphasized rapid heating of dilute coal samples which has been
amply termed "flash pyrolysis". Behind this shift in emphasis have been the de-
sire to elucidate the role of pyrolysis in combustion and gasification and the
possibility of applying pyrolysis to convert coal to liquid fuels or chemicals.
The material in this monograph is mainly addressed to pyrolysis under conditions
of rapid heating.

Chapters 2 and 3 are devoted to the chemistry of pyrolysis. In Chapter 2,
recent spectroscopic and wet chemical information is utilized to describe coal's
structure in terms of functional groups. Since most of the experimental data
concern coal derived liquids, extrapolation to solid coal involves considerable
uncertainty. However tentative it may be, this extrapolation is necessary for
mechanistic analysis of pyrolysis. Having established or postulated the pertinent
functional groups, it is possible to identify with considerable confidence the
important elementary thermal reactions. Basic organic chemistry and kinetic
studies with model compounds, provide a firm basis for this purpose. Chapter 3
contains a survey of the energetically favored elementary reactions and discusses
techniques for estimating their rate parameters. The estimation techniques are

illustrated with numerous examples. However, a number of elementary reactions defy estimation by these established predictive techniques. This is the case, for example, with reactions which are limited by diffusion in the condensed phase.

The bulk of experimental results are presented in Chapter 4. The data consist of the yield of various pyrolysis products as a function of temperature and pyrolysis time under conditions that minimize the influence of heat and mass transfer limitations. A brief discussion of commercial process development efforts is also included in this chapter. Chapter 5 is devoted specifically to heat and mass transfer limitations as manifested by the dependence of product yields on pressure and particle size. Although the chapter contains experimental and modeling results, the modeling part is limited to delineating conditions under which transport phenomena become important.

In the presentation of experimental results in chapters 4 and 5, a distinction is made between softening coals, primarily high volatile bituminous, and nonsoftening coals e.g. subbituminous or lignites. This rudimentary distinction does not do justice to the variety of composition and properties within each of the two categories. However, it is a point of departure and has particular significance relative to the evolution of porous structure and plastic properties both of which strongly influence mass transfer rates.

Chapter 6 contains a survey of kinetic models of pyrolysis with emphasis on underlying assumptions rather than mathematical details. A brief discussion is also given about the ability of each model to describe experimental data. This ability does not by itself imply the validity of the underlying assumptions. Product yields are smooth and gently sloping functions of operating conditions, that could be fitted by many phenomenological models with reasonable kinetic form, provided the rate parameters are in each case suitably adjusted. Quite often the scatter in the data masks subtle distinctions between alternative models. Phenomenological models have reached a maturity that allows their application to coal combustion or gasification. By contrast, the one or two detailed chemical models discussed in this chapter are still at a stage of flux and are not ripe for practical applications.

The final Chapter, 7, contains a brief survey of experimental data and modeling work on hydropyrolysis, the heating of coal in a hydrogen atmosphere. This variant of pyrolysis is receiving increased attention as one of the most promising routes to chemicals from coal.

Chapter 2

CHEMICAL STRUCTURE OF COALS

2.1   FUNCTIONAL GROUPS
   The scope of any systematic description of coal structure is seriously limited
by the large variety of coals and the variability in the chemical properties of
different macerals in a single coal. Nevertheless, such a description becomes
meaningful if it is postulated that the chemical properties of coal are determined
by a modest number of functional groups common to all coals. Coals are then assumed
to differ by their content of these functional groups. Although different macerals
(vitrinite, fusinite, exinite, etc.) differ markedly in composition and reactivity,
in this treatise we will not take into account such differences, since most of
the pyrolysis data till now have been obtained with whole coals. However, in as
much as vitrinite is the dominant maceral (more than 70% in most coals), the
structural features discussed below essentially refer to the vitrinite component.
The idea of representing different coals by a common set of functional groups has
not been fully justified but is generally compatible with the experimental evidence
discussed in the next section.
   The reactivity of coal in pyrolysis, hydropyrolysis and liquefaction can be
characterized by six classes of functional groups: aromatic nuclei, hydroaromatic
structures, alkyl chains, alkyl bridges and oxygen groups. Sulfur and nitrogen
groups are important from the standpoint of pollutant formation but do not play
a significant role in the thermal reaction mechanisms, if only due to their generally low content. Each of these classes of functional groups will be discussed
below drawing mainly from the recent literature.

2.1.1 Aromatic nuclei

   Nuclear magnetic resonance spectroscopy (^{1}H and ^{13}C) of coals and coal liquids
has established that 40-75% of carbon in subbituminous and bituminous coals is
aromatic, the aromaticity increasing with rank. The ability to determine the
aromaticity reasonably accurately has blunted or made irrelevant earlier arguments
about the nature of the carbon skeleton in coal. Aromatic rings of various sizes
are important building blocks of coal structure and maintain their integrity when
coal is heated at temperatures as high as 700°C. Even when the aromatic content
is as low as 40%, the reactivity of the aliphatic carbon is strongly influenced
by proximal aromatic nuclei.
   A variety of techniques have been employed to determine the size of the aromatic nuclei. Early studies employing X-ray diffraction spectroscopy, reviewed
by Speight and Given (refs. 1,2) suggested that coals with up to 87% carbon (this

includes lignites, subbituminous and bituminous coals) contain mainly units with
3 to 5 condensed rings and a smaller but appreciable amount of single and double
rings. This conclusion is subject to considerable uncertainty in view of a number of assumptions that had to be made concerning the geometric arrangement and
the disposition of aliphatic substituents. Fluorescence spectroscopy of bituminous
coal extracts (ref. 3) and comparisons with model compounds suggest that the
more abundant aromatic nuclei consist of three or more condensed rings. Polaro-
graphic analysis of SRC liquids of a high volatile C bituminous coal has indicated
a small content of units with more than two condensed rings (ref. 4). Ruberto,
et al. (ref. 5) applied a variety of techniques including ^{1}H nmr, U.V. and mass
spectroscopy to a solvent extracted subbituminous coal to conclude that the aromatic units consist of two and three condensed rings. Since these analyses were
performed on coal liquids, drawing inferences about the structure of solid coal
involves some uncertainty.

The size of the aromatic nuclei can also be estimated from the elemental composition and ^{1}H nmr spectra of coal extracts or other coal derived liquids by a procedure known as "structural analysis", described in the next section. In as much
as structural analysis employs several assumptions, these estimates are subject
to additional uncertainty. Tingey and Morrey (ref. 6) have estimated an average
size of 5-8 condensed rings for the extracts of a subbituminous and two bituminous
coals. Whitehurst (ref. 4) has investigated the asphaltol fraction of a liquefied
bituminous coal and proposed structures consisting of one, two and three condensed
rings. Gavalas and Oka (ref. 7) have applied the Brown and Ladner structural
analysis (see next section) to the extract of a bituminous coal to estimate nucleus
size of two to three condensed rings.

The consensus emerging from the aforementioned studies is that lignites and
subbituminous coals contain nuclei with one to three condensed rings, while bituminous coals contain mostly two to four condensed rings, the degree of condensation increasing with the coal rank.

Two and three condensed ring nuclei found in substituted form in coal derived
liquids include naphthalene, indene, biphenyl, furan, pyrrole, pyridine, benzo-
thiophene, phenanthrene, anthracene and various three-ring heterocycles of oxygen,
sulfur and nitrogen. Some of the rings may be partly hydrogenated constituting
what are known as "hydroaromatic" structures.

2.1.2 Aliphatic structures

The aliphatic segments of coal include aliphatic chains (methyl, ethyl, etc.),
bridges (methylene, ethylene) and hydroaromatic structures. The information
available about aliphatic structures derives mainly from proton and ^{13}C magnetic
resonance spectra and from selective oxidation, especially by the recently developed technique of Deno and coworkers (refs. 12-14).

Proton and ^{13}C magnetic resonance has been fruitfully applied to coal derived liquids: pyrolyzates, extracts, and hydrogenation products that are all modified fragments of the original coal. The modification is relatively small in the extraction, or H-donor mediated extraction at low conversions, somewhat larger in flash pyrolysis, and rather severe in liquid phase hydrogenation to distillate products.

Figure 2.1. below illustrates the information derived from nuclear magnetic resonance. The ^{1}H nmr spectrum distinguishes three major types of hydrogen:

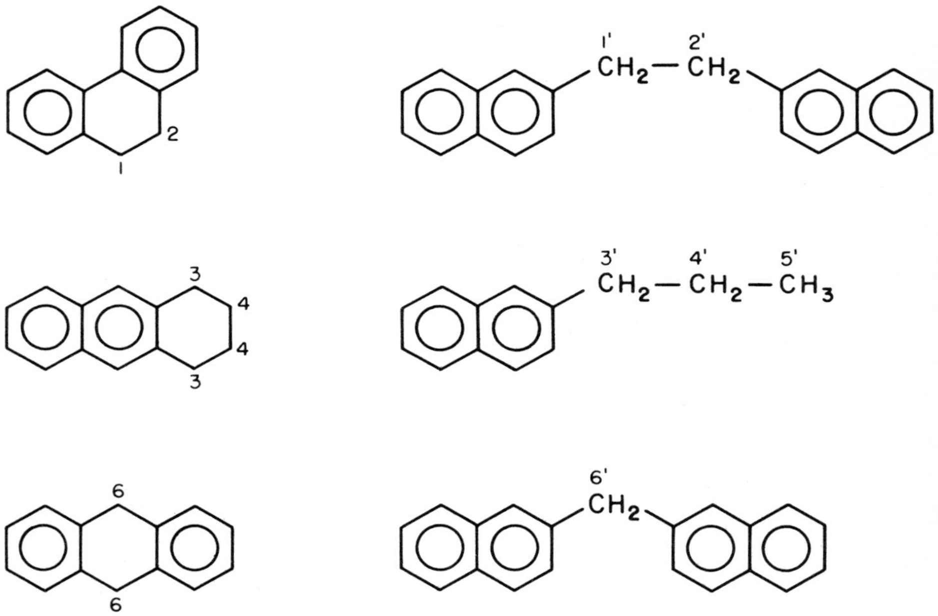

Fig. 2.1. Structural information derived from nmr spectra.

$H_{ar + o}$, aromatic and phenolic hydrogen, H_α, hydrogen at position α as 1,2,1´, 2´,3,3´,6,6´ of Figure 2.1 and $H_{\beta+}$, hydrogen at position β and further from the ring as 4,4´,5´ of Fig. 2.1. In recent years it has been possible to further resolve the alpha and beta region. For example, Bartle et al.(refs. 8,9) distinguished two types of α and several types of β hydrogen in the ^{1}H nmr spectra of supercritical coal extracts. Similar distinctions were made in the spectra of coal hydrogenation liquids by Yokohama et al. (ref. 10). Although high resolution ^{1}H nmr spectra provides very detailed information on hydrogen types, the interpretation of the spectra is not without some ambiguity. For example, the distinction between the hydrogens 1´ and 3 in Fig. 2.1 is somewhat uncertain.

^{13}C nmr spectra have been similarly used to distinguish among C_{ar}, and several types of C_α and C_β (refs. 9, 11), however, drawing quantitative results from these spectra involves a larger error than in the case of ^1H spectra.

Deno and coworkers (refs. 12-14) demonstrated with model compound studies summarized in Table 2.1 that an aqueous solution of H_2O_2 and trifluoroacetic acid selectively oxidizes the aromatic part of a molecule, leaving much of the aliphatic part intact. Among the model compounds listed in Table 2.1 only D gives tri- and tetra-carboxylic acids.   Thus, the presence of the latter acids in the oxidation products of the Illinois #6 coal was regarded as evidence that D or similar structures constitute a large fraction of the structural units of this coal. Unfortunately, no work was performed with compounds like

to assess the possible role of ethylene bridges. The distinction between methylene and ethylene bridges is important from the standpoint of reactivity as will be discussed in a subsequent chapter.

As shown in Table 2.1, the oxidation products from the subbituminous Wyodak coal are the same as from the Illinois #6 but the relative amounts are very different. The Wyodak coal gave a lower yield of total products and a very small yield of polycarboxylic acids. These differences are probably due to the smaller ring structures and the larger concentration of phenolic substituents in the subbituminous coal. Both coals gave a little acetic acid, evidently from methyl substituents, considerable malonic acid, presumably from methylene bridges, and considerable succinic acid. The latter acid could be due to either ethylene bridges (diarylethane structures) or 9,10-dihydrophenanthrene. It should be noted that the compounds identified in the oxidation products of tetralin are essentially absent from both coals indicating negligible participation of tetralin type structures.   The results obtained by this technique are intriguing but additional model compound work is needed to reach definitive conclusions.

2.1.3 Oxygen functionalities

Oxygen plays a pervasive role in almost all aspects of coal reactivity. Nitrogen, in the form of basic groups, is involved in the secondary intermolecular forces in coal or coal derived liquids. It is also the source of much of the NO_x generated in coal combustion.   Sulfur is of concern primarily as a pollutant precursor.

Oxygen is present as phenolic hydroxyl and carboxylic acid groups, aryl-aryl or alkyl-aryl ether bridges, and ring oxygen, chiefly in furan-type structures. Phenolic hydroxyls are evident in infrared spectra (refs. 15,16) and have been

TABLE 2.1

Products from the oxidation of pure compounds with aqueous $H_2O_2+CF_3COOH$ (refs. 12-14).

Compound oxidized	Yields of carboxylic acids (weight %)				
	Acetic	Propionic	Malonic	Succinic	Other
toluene	87				
ethylbenzene	19	71			
propylbenzene	5	8			butyric: 73
2-methylnaphthalene	83				
diphenylmethane			9		
diphenylethane			9	73	
tetralin					cyclohexene-1,2-dicarboxylic anhydride: 71
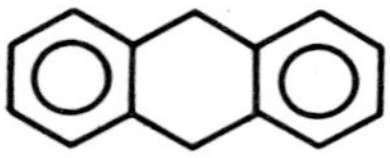				14	benzenedicarboxylic acid: 70
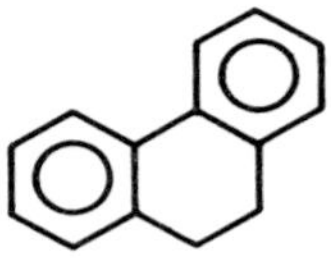				31	benzenedicarboxylic acid: 9, other
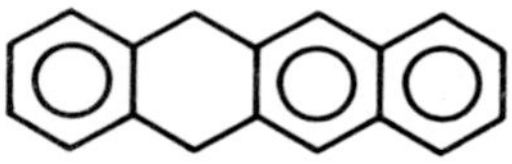 D			23[a]		benzenemono-, di-, tri- and tetracarboxylic acids: 200[a]
Illinois #6 bituminous	1.4[b]		2.4[b]	7.4[b]	same as above: 30[b]
Wyodak subbituminous	1.8[c]		11.4[c]	4.9[c]	same as above: 7[c]

a, b, c = relative yields

quantitatively determined by a technique based on the formation of trimethylsilyl
ethers (ref. 17). They constitute one to two thirds of the oxygen in coal. The
determination of **ether** bridges which occur in smaller amounts has been indirect
and less reliable. Some investigators (e.g. ref. 18) have suggested that ether
bridges are the predominant links between the structural units of bituminous
coals. However, the bulk of the evidence is that aliphatic bridges are at least
as abundant. Carboxylic groups are found in considerable concentration in subbituminous coals and lignites. Furan-type ring oxygen comprises a small fraction
of the total and is relatively unimportant from the standpoint of reactivity. A
quantitative determination of oxygen functional groups in several coals has been
made in connection to liquefaction studies (ref. 19).

The functional groups discussed above are organized in *modular units* (or simply
units) which are covalently linked to *coal molecules*. The coal molecule in Fig.
2.2 is shown to illustrate terminology rather than to suggest a representative
structure. We do not wish to join in the debate surrounding "coal models". A
modular unit, e.g. unit A, consists of an aromatic *nucleus,* in this case naphthalene, aliphatic chains, mostly methyl or ethyl, phenolic hydroxyl and carboxylic
acid substitutents. Some units such as unit B contain a partially hydrogenated
nucleus, in this case the 9,10-dihydrophenanthrene. The modular units are linked
by *bridges,* mostly methylene, ethylene and ether. The term bridge will denote a
link that can be thermally broken. The unit C, for example, contains a biphenyl
nucleus. The bond between the two benzene rings is not counted as a bridge because of its high dissociation energy.

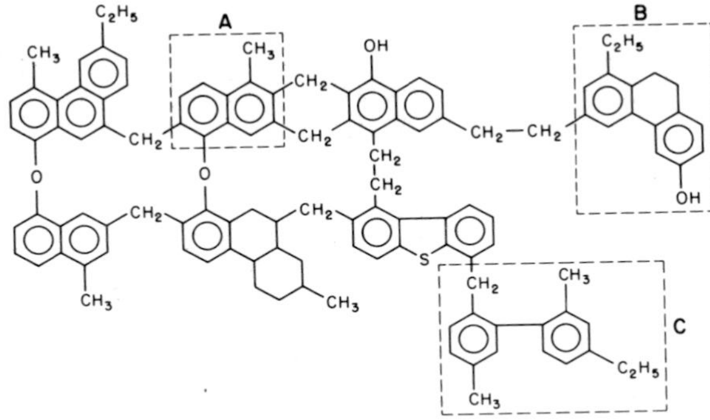

Fig. 2.2. Illustration of modular units, nuclei, bridges and chains in a coal
molecule.

It has been indicated earlier that the most abundant nuclei in subbituminous and bituminous coals consist of two or three condensed rings each, as illustrated in Fig. 2.2. It may be further inquired what is the number of modular units or the molecular weight of the coal molecule. A seasoned discussion of the evidence concerning molecular weights was presented by Given (ref. 2). The number average molecular weight of pyridine extracts of coal and of a Li/amine reduced coal (reducing aromatic rings but probably not cleaving bonds) were in the range 1,000 to 4,000. The extracts of the reduced coal represented 40-80% of the original coal by weight, hence the molecular weight of 1,000-4,000 can be considered fairly representative. The remaining insoluble material may have higher molecular weight or higher concentration of polar groups.

Assuming a molecular weight of the modular unit of about 200-250, the coal molecule should contain 5 to 20 modular units. The arrangement of these units is unlikely to be linear, because that would imply a number of bridges much lower than inferred from devolatilization experiments. A two-dimensional structure comprising sheets is certainly possible. The coal molecules are held together by secondary (non-covalent) bonds like hydrogen bonds or charge transfer acid-base complexes. Such secondary bonds are severed during extraction by pyridine or other good solvents or at elevated temperatures.

2.2   STRUCTURAL AND FUNCTIONAL GROUP ANALYSIS

The analytical data that are commonly available for coal derived liquids - extracts, pyrolyzates, and products of hydrogenation - are elemental analysis, and ^1H nmr spectra and, less routinely, ^{13}C spectra. These data can be translated to a form easy to visualize and correlate with reactivity properties by certain procedures that have become known as structural analysis and functional group analysis. The more common method of structural analysis seeks to determine features such as the size of the aromatic nuclei, the degree of substitution and the relative abundance of various types of aliphatic chains. If the molecular weight is also specified, perhaps in some narrow range, representative or "average" molecular formulas can be constructed compatible with the data. The related term functional group analysis indicates the determination of the concentration of specific funtional groups without being concerned with representative molecular formulas.

2.2.1 The Brown and Ladner method (refs. 20,21)

This simplest and best established of the structural analysis procedures utilizes the atomic ratios C/H, O/H, and the hydrogen distribution H_{ar+o}/H, H_{α}/H, H_{β}/H, i.e. the fractions of aromatic + phenolic hydrogen, alpha hydrogen and beta or further removed from the aromatic ring hydrogen. The following assumptions or specifications are made:

(i)   A certain fraction (usually 60%) of the oxygen is assumed to be phenolic in accordance with wet chemical studies indicating a phenolic fraction of oxygen in the range 0.4 - 0.6.

(ii) The remaining oxygen (other than phenolic) is assumed to be connected to aromatic carbons as substituent (e.g. quinonic structures or methoxy groups), and not as ring oxygen.

(iii) The ratios $x = H_\alpha/C_\alpha$, $y = H_\beta/C_\beta$ are specified within reasonable limits. The values $x = y = 2$ have been used in the paper of Brown and Ladner (ref. 21).

(iv) Biphenyl bonds are assumed absent.

(v)   Sulfur and nitrogen groups are ignored.

Subtracting the phenolic hydrogen (assumption ii) from H_{ar+o} provides the aromatic hydrogen, H_{ar}/H.   The aromaticity f_a, i.e. the fraction of aromatic carbon, is then calculated with the aid of assumption (iii),

$$f = \frac{C - C_\alpha - C_\beta}{C} = \frac{(C/H) - (H_\alpha/xH) - (H_\beta/yH)}{(C/H)}$$

The next quantity of interest is the degree of substitution σ, which is the ratio of substituted aromatic carbons to the total number of peripheral aromatic carbons.   This is calculated using assumptions (iv) and (v):

$$\sigma = \frac{C_\alpha + O}{C_\alpha + O + H_{ar}} = \frac{(H_\alpha/xH) + (O/H)}{(H_\alpha/xH) + (O/H) + (H_{ar}/H)}$$

Finally, if H_{aru} is the number of aromatic hydrogen and other aromatic substituents and C_{ar} the number of aromatic carbons, the quantity H_{aru}/C_{ar} which represents the ratio of H and C in the hypothetical unsubstituted hydrocarbon, is given by:

$$\frac{H_{aru}}{C_{ar}} = \frac{H_{ar} + C_\alpha + O}{C_{ar}} = \frac{(H_{ar}/H) + (H_\alpha/xH) + (O/H)}{(f_a C/H)}$$

In addition to its indirect determination given above, the aromaticity can be obtained directly from ^{13}C magnetic resonance spectra.   The two values were found in very good agreement in a study of various coal derived liquids (ref. 22).

The ratio H_{aru}/C_{ar} is related to the size of the aromatic nucleus.   For example, naphthalene, anthracene and phenanthrene have H_{aru}/C_{ar} values of 0.8, 0.714 respectively.   A computed value of 0.75 might thus indicate a mixture largely consisting of naphthalene and phenanthrene or anthracene.   This interpretation of H_{aru}/C_{ar} is obscured, however, by the presence of heteroatoms, biphenyl bonds and hydroaromatic nuclei such as 9,10-dihydrophenanthrene.

2.2.2 Detailed structural analysis (refs. 23,24)

Bartle and Smith (ref. 23) developed a detailed characterization of the aliphatic structures based on the ability to break down the ^1H nmr spectrum into the following types of hydrogen: H_{ar+o}, hydrogen in methylene connecting two rings (e.g. a methylene bridge); hydrogen of methylene in acenaphthene or indene as well as methylene connected to rings in highly sterically hindered positions; H_α not belonging to the two previous categories; hydrogen in -CH$_2$-, -CH- groups β or further to a ring + CH$_3$ β to a ring; hydrogen of CH$_3$ γ or further to a ring. These data are supplemented by the following assumptions:

(i)   A certain fraction of oxygen is attributed to phenolic groups; the remaining is assumed to be in dibenzofurans.

(ii)   Nitrogen is assumed to be in rings of the type of pyridine and carbazole, while sulfur is assumed to be in dibenzothiophene rings.

(iii) The aliphatic chains considered include methyl, ethyl, propyl, isopropyl, n-butyl, isobutyl and sec-butyl. The ratio of n-butyl to n-propyl is given some convenient value.   The numbers of n-butyl, i-butyl and sec-butyl are taken to be equal.

Using assumptions (i) and (ii)the heteroatoms are replaced with C and H to produce an equivalent structure containing C and H only.   Balances are then written for H_α, H_β and H which in conjunction with assumption (iii) yield the concentrations of the various aliphatic chains.

The above procedure has been used to determine *representative* structures for narrow molecular weight fractions of a supercritical coal extract (ref. 8). For example, the structure below was proposed for the methanol eluate of the extract:

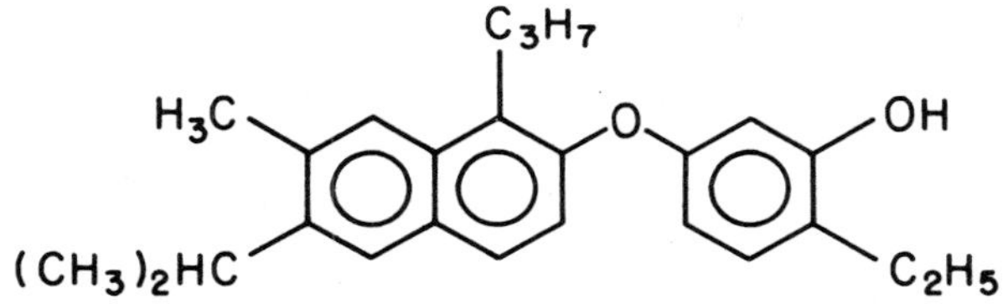

The method was recently modified and further applied to characterize coal extracts (ref. 24).   By combining the information from ^{1}H and ^{13}C nmr spectra, it was possible to largely dispense with the assumptions (i)-(iii) above in an application to petroleum fractions (ref. 25). The fundamentals of ^{1}H and ^{13}C nmr spectroscopy as it pertains to complex mixtures, coal liquids in particular, has been recently reviewed by Bartle and Jones (ref. 26).

2.2.3 Computer-assisted molecular structure construction (ref. 27)

The structural parameters and the concentrations of functional groups derived by the methods of refs. 25-26 can be assembled into average or representative molecular formulas, assuming suitable molecular weight information. In the aforementioned references the construction of molecular formulas was carried out by trial and error. Since this procedure becomes impractical with increasing molecular weight, Oka et al.(ref. 27) developed a computer-based technique for generating molecular formulas from data on ^1H nmr, elemental analysis and molecular weight.

2.2.4 Functional group analysis by linear programming

As already discussed, structural analysis can be carried out at two levels. At the first level, it is sought to derive various features such as f_a or H_{aru}/C_{ar} representing averages for the mixture examined. The concentrations of various aliphatic groups can also be estimated if the ^1H nmr data are sufficiently detailed. At the second level, a representative or average moelecular formula is sought, compatible with the available data. Such a formula is illustrative of the type of possible structures but cannot convey the variety of different molecules that may be present even in a narrow fraction of the sample.

The thermal reactions of pyrolysis, hydropyrolysis and liquefaction can be largely understood in terms of functional groups rather than molecules, therefore the first level of structural analysis is usually sufficient. The problem can then be stated as follows: Given elemental analysis, nmr and possibly other data calculate the concentrations of a postulated set of functional groups. Invariably, the number of unknowns exceeds the number of balance equations incorporating the data, therefore the problem accepts an infinite number of solutions. To obtain a unique solution, additional assumptions in the form of values for the ratios of some of the unknowns may be introduced as in refs. 23-25. This is a convenient device but in some cases the specified values may lead to improbable, even negative values for some concentrations. Thus, some trial and error is normally required in the choice of the assumed values.

A more systematic procedure for generating meaningful solutions of the undetermined problem of functional group analysis is afforded by linear programming. Suppose that $y_1,\ldots,y_n$ are the unknown concentrations of the selected functional groups and let the balance equations have the form

$$\sum_{j=1}^{n} A_{ij}y_j = b_i \qquad i = 1,\ldots,m \qquad (2.1)$$

where $m<n$. In the linear programming formulation it is required to maximize a quantity

$$J = q_1y_1 + \ldots + q_ny_n \qquad (2.2)$$

subject to Eq. (1.1) and the obvious constraint $y_i \geq 0$. In this notation A_{ij} are stoichiometric coefficients, b_i are conserved quantities, directly derived from the data, and q_i are positive numbers chosen on the basis of experience and intuition. Suppose, for example, that y_1,y_2 represent the concentrations of methylene and ethylene bridges. A complete lack of preference as to which of the two is more abundant can be expressed by setting $q_2=2q_1$, the factor 2 deriving from the two-to-one carbon between the two bridges. To express an intuitive preference for methylene bridges, on the other hand, one might set $q_2 = 0.2q_1$

For purposes of illustration this technique was applied to a product of lique-faction of a bituminous coal with the analytical data given in the following table.

TABLE 2.2
Data for a donor solvent extract of a bituminous coal (ref. 28).

Conserved quantity	Concentration 10^{-2} g-mols/g extract	Conserved quantity	Concentration 10^{-2} g-mols/g extract
C_{ar}	$b_1 = 5.31$	$H_{\beta+}$	$b_6 = 0.52$
C_{al}	$b_2 = 1.86$	0	$b_7 = 0.28$
H_{ar+o}	$b_3 = 2.14$	S	$b_8 = 0.16$
H_{α}	$b_4 = 2.21$	N	$b_9 = 0.14$
H_{β}	$b_5 = 1.62$		

The functional groups used in this illustration, twenty-one in number, are listed in Fig. 2.3. The coefficients A_{ij} are obtained by inspection of Table 2.2 and Fig. 2.3. A_{ij} is the contribution of functional group j (Fig. 2.3) to the conserved quantity i (Table 2.2). For example, $A_{2,13} = 2$, $A_{3,13} = -1$ because a chain $-CH_2CH_3$ replaces a group H_{ar+o} which would otherwise be attached to the aromatic carbon. Similarly, $A_{3,17} = -2$, $A_{5,16} = -1$ because the group $o-CH_3$ re-places a H_{β}. Table 2.4 later gives the complete matrix of coefficients A_{ij}.

As mentioned below, the choice of the coefficients q_i is a statement of experi-ence or intuition concerning the relative abundance of various functional groups. For example $q_1 = 1$, $q_2 = 5$, $q_3 = 5$, $q_4 = 1$, $q_5 = 1$ states that it is more prob-able that the mixture contains groups 2 and 3 than groups 1, 4 and 5. The re-sulting solution to the problem may still contain groups 1, 4 and 5, however. Even when $q_j = 0$, the solution may still contain $y_j > 0$. Each q-set provides a solution satisfying equations (2.1). Different q-sets do not necessarily pro-duce different solutions. If $y_1, \ldots, y_n$ and $y'_1, \ldots, y'_n$ are the solutions corres-ponding to $q_1, \ldots, q_n$ and $q'_1, \ldots q'_n$ then the linear combination $y''_j = \lambda y_j + (1-\lambda)y'_j$ ($0 \le \lambda \le 1$) is also a solution of equation (2.1) although it does not maximize any particular J. However, it is perfectly acceptable from our standpoint. The util-ization of the maximization problem is mainly a device to obtain feasible solu-tions.

14

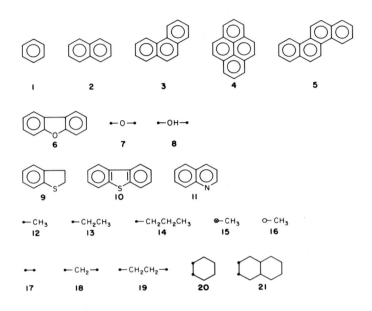

Fig. 2.3. Functional groups used in the analysis of the data of Table 2.2. The symbols ●, ⊗, ○ denote attachment to aromatic, α and β carbons, respectively.

Table 2.4 lists eleven distinct q-sets chosen to emphasize different functional groups. For example, sets 3 and 4 differ only in the coefficients q_2, with set 3 emphasizing the phenathrene ring while set 2 placing equal emphasis on naphthalene and phenanthrene. Table 2.5 gives the results of calculations with each of the q-sets of Table 2.4. Several features of these results should be noted:

(i) For any given set of q_j, the number of non-zero y_j is equal to the number of data points, m. The remaining y_j are zero. By combining such solutions it is possible to generate feasible solutions with a larger number of positive components. All the q-sets of Table 2.4 produce zero values for the components y_4, y_5, y_6, y_{10}, y_{14}, y_{20}, y_{21} without precluding, however, that some other q-set may produce non-zero values for some of these components.

(ii) A comparison of the solutions corresponding to the q-sets 5 and 6 differing only in q_1, q_2, and q_3 shows set 5 producing only benzene and set 6 producing only naphthalene. The transition from benzene to naphthalene is accompanied by a reduction in the number of biphenyl bridges.

TABLE 2.3
Matrix of coefficients A_{ij}

j i	1	2	3	4	5	6	7	8	9
1	6	0	6	0	0	0	0	0	0
2	10	0	8	0	0	0	0	0	0
3	14	0	10	0	0	0	0	0	0
4	16	0	10	0	0	0	0	0	0
5	18	0	12	0	0	0	0	0	0
6	12	0	8	0	0	0	1	0	0
7	0	0	-2	0	0	0	1	0	0
8	0	0	0	0	0	0	1	0	0
9	8	0	6	0	0	0	0	1	0
10	12	0	8	0	0	0	0	1	0
11	9	0	7	0	0	0	0	0	1
12	0	1	-1	3	0	0	0	0	0
13	0	2	-1	2	3	0	0	0	0
14	0	3	-1	2	2	3	0	0	0
15	0	1	0	-1	3	0	0	0	0
16	0	1	0	0	-1	3	0	0	0
17	0	0	-2	0	0	0	0	0	0
18	0	1	-2	2	0	0	0	0	0
19	0	4	-2	4	4	0	0	0	0
20	0	4	-2	4	4	0	0	0	0
21	0	8	-2	4	2	8	0	0	0

(iii) The concentrations of the hydroaromatic groups 20,21 are zero for all q-sets. It should be kept in mind, however, that in the functional group analysis a structure like

is regarded as containing two methylene bridges rather than a hydroaromatic group.
(iv) The q-sets with $q_7 = q_8$ yield concentrations $y_7 = 0$, $y_8 > 0$. After sufficiently increasing q_7 the concentrations become $y_7 > y_8 = 0$ with a simultaneous change in the functional groups 17,18 and 19 representing bridges. A solution with $y_7 > 0$, $y_8 > 0$ can be obtained by combining the solutions of q-sets 6 and 7 using equal weights. This solution has 11 positive components and is listed under column C of Table 2.5.

TABLE 2.4

Sets of coefficients q_i

	Set. No.										
	1	2	3	4	5	6	7	8	9	10	11
q_1	2	1	1	1	2	1	1	1	1	1	1
q_2	2	1	5	1	2	5	5	1	1	1	1
q_3	2	1	5	5	2	5	5	5	20	20	20
q_4	1	1	1	1	1	1	1	1	1	1	1
q_5	1	1	1	1	1	1	1	1	1	1	1
q_6	2	1	2	2	2	2	2	2	2	2	2
q_7	4	1	4	4	4	4	5	5	4	4	4
q_8	4	1	4	4	4	4	2	2	4	4	4
q_9	5	1	5	5	5	5	5	5	5	5	5
q_{10}	5	1	5	5	5	5	5	5	5	5	5
q_{11}	10	1	10	10	10	10	10	10	10	10	10
q_{12}	2	1	2	2	5	5	5	5	2	2	2
q_{13}	2	1	2	2	5	5	5	5	2	2	5
q_{14}	2	1	2	2	5	5	5	5	2	2	2
q_{15}	2	1	2	2	5	5	5	5	2	2	2
q_{16}	2	1	2	2	5	5	5	5	2	2	2
q_{17}	2	1	2	2	2	2	2	2	2	1	1
q_{18}	4	1	4	4	2	2	2	2	4	1	1
q_{19}	2	1	2	2	2	2	2	2	2	4	4
q_{20}	5	1	5	5	5	5	5	5	5	5	5
q_{21}	5	1	5	5	5	5	5	5	5	5	5

TABLE 2.5
Concentrations of functional groups (10^{-2} g-mols/gr extract) computed by
linear programming for various q sets.

Group No.	q-set									
	1	2	3	4	5	6	7	8	9	C
1	6.59	6.59	0	6.59	6.59	0	0	6.59	0	0
2	0	0	3.95	0	0	3.95	3.95	0	0	3.95
3	0	0	0	0	0	0	0	0	2.82	0
4	0	0	0	0	0	0	0	0	0	0
5	0	0	0	0	0	0	0	0	0	0
6	0	0	0	0	0	0	0	0	0	0
7	0	0	0	0	0	0	2.82	2.82	0	1.41
8	2.82	2.82	2.82	2.82	2.82	2.82	0	0	2.82	1.41
9	0.16	0.16	0.16	0.16	0.16	0.16	0.16	0.16	0.16	0.16
10	0	0	0	0	0	0	0	0	0	0
11	1.36	1.36	1.36	1.36	1.36	1.36	1.36	1.36	1.36	1.36
12	6.21	6.21	6.21	6.21	6.21	6.21	6.21	6.21	6.21	6.21
13	0	0	0	0	0	0	0	0	0	0
14	0	0	0	0	0	0	0	0	0	0
15	5.99	5.99	5.99	5.99	5.99	5.99	5.99	5.99	5.99	5.99
16	1.73	1.73	1.73	1.73	1.73	1.73	1.73	1.73	1.73	1.73
17	6.48	8.84	2.53	6.48	6.48	2.53	0	3.66	0.83	1.26
18	4.71	0	4.71	4.71	4.71	4.71	4.13	4.71	4.71	4.42
19	0	2.36	0	0	0	0	0.29	0	0	0.15
20	0	0	0	0	0	0	0	0	0	0
21	0	0	0	0	0	0	0	0	0	0

C: average of results of q-sets 6 and 7.

Chapter 3

THERMAL REACTIONS OF COAL

This chapter deals with the rate parameters of the principal elementary reactions of pyrolysis, hydropyrolysis and liquid-phase hydrogenation. While considerable data is available in the literature about rate parameters, in most cases recourse must be made to estimation techniques such as group additivity and transition state theory. These two techniques, group additivity in particular, have been largely developed by Benson and coworkers (e.g. refs. 29,30) and are popularly known as thermochemical kinetics. Much of this chapter is taken up by the application of thermochemical kinetics to various elementary reactions of interest. The estimated parameters along with those taken from the literature are compiled in a table at the end of the chapter.

One of the uncertainties associated with estimating rate parameters for coal reactions is due to the condensed nature of the coal phase which is the reaction medium. Almost all reported rate values and group contribution information refer to gas phase kinetics. Another uncertainty arises from the fact that coal molecules contain, primarily, condensed ring aromatics while much of the reported experimental information refers to single ring compounds. When these uncertainties are added to the significant uncertainty associated with the structure of the transition state and the frequencies of its vibrational modes it will be realized that estimated A-factors could be easily in error by a factor of ten. Nevertheless, the estimates discussed here are very useful in the absence of other information and can obviously be improved by comparison with data from model compound studies. With these difficulties in mind we now proceed to examine several classes of reactions: bond dissociation, hydrogen abstraction and hydrogen addition. These are all free radical reactions. Reactions other than free radical, e.g. concerted reactions, have only recently been examined relative to coal pyrolysis and will be briefly surveyed in the last section.

3.1 BOND DISSOCIATION WITH PRODUCTION OF TWO RADICALS

The following reactions will be considered with Ph, Ph´ representing aromatic nuclei such as naphthalene, phenanthrene, etc.

$$Ph-CH_3 \rightarrow Ph-CH_2^{\cdot} + H^{\cdot} \tag{D1}$$
$$Ph-CH_2CH_3 \rightarrow Ph-CH_2^{\cdot} + CH_3^{\cdot} \tag{D2}$$
$$Ph-CH_2CH_2CH_3 \rightarrow PhCH_2^{\cdot} + CH_3CH_2^{\cdot} \tag{D3}$$
$$Ph-CH_2-Ph´ \rightarrow Ph-CH_2^{\cdot} + Ph´^{\cdot} \tag{D4}$$
$$Ph-CH_2CH_2-Ph´ \rightarrow Ph-CH_2^{\cdot} + Ph´-CH_2^{\cdot} \tag{D5}$$
$$Ph-O-CH_3 \rightarrow Ph-O^{\cdot} + CH_3^{\cdot} \tag{D6}$$

Ph-O-Ph´ → Ph-O· + Ph´·   (D7)

Ph-O-CH$_2$-Ph´ → Ph-O· + Ph´-CH$_2$·   (D8)

   The rate of each of these elementary reactions can be expressed in the Arrhenius form

$k = A\exp(-E/RT)$

where A is the "A-factor" and E the experimental activation energy, both being functions of temperature. The activation energy E can be estimated using the fact that the reverse reaction, free radical recombination, has very small acti- vation energy which is conventionally taken as zero. Thus

$E = \Delta H$ (3.1)

where ΔH is the standard heat (or enthalpy) of reaction, at the temperature of interest and pressure of one atmosphere. ΔH is also somewhat less accurately known as the bond dissociation "energy". The heat of reaction ΔH can be estimated by the group additivity method as shown in the examples below.

   The A-factor is given by transition state theory as

$$A = e \; \frac{k_B T}{h} \; \exp(\Delta S\circ^{\dagger}/R)$$ (3.2)

where $\Delta S\circ^{\dagger}$ is the standard (atmospheric pressure) activation entropy, i.e. the difference in entropy between the transition state (or activated complex) and the reactants. The activation entropy can be estimated by the methods of statis- tical mechanics if the structure of the activated complex is known. This estima- tion involves a great deal of uncertainty because of lack of reliable information about several of the vibrational frequencies of the activated complex.

   Benson and O'Neal (ref. 29) have evaluated experimental data for several of the reactions (D1)-(D8) and have recommended preferred parameter values in each case. In the following subsections we shall illustrate the theoretical estimation tech- niques on reactions which are not included in the compedium of Benson and O'Neal. The estimated parameters along with values from the literature are listed in the table at the end of the chapter.

   Examples 1 to 5 treat the estimation of activation energies while examples 6 to 8 the estimation of A-factors.

3.1.1 Activation energies

Example 1.

   In this example we examine reaction D1 with Ph an unsubstituted phenyl. The calculations will be initially made for 300°K since most of the available data refer to that temperature. Subsequently, the estimate will be revised to apply to 800°K. The heat of reaction can be expressed in terms of heats of formation as

$\Delta H = \Delta_f H(H\cdot) + \Delta_f H(Ph\text{-}CH_2^{\cdot}) - \Delta_f H(Ph\text{-}CH_3)$

The last two terms can be estimated by group additivity as follows:

$\Delta_f H(Ph-CH_2^{\cdot}) = 5[C_B-(H)] + [C_B-(C^{\cdot})] + [C^{\cdot}-(C_B)(H)_2]$

$\Delta_f H(Ph-CH_3) = 5[C_B-(H)] + [C_B-(C)] + [C-(C_B)(H)_3]$

where the brackets denote heats of formation and the groups within the brackets are written with the notation of ref. 30. The tables in ref. 30 give $[C_B-(C^{\cdot})] = [C_B-(C)]$, hence

$\Delta H = [H^{\bullet}] + [C^{\cdot}-(C_B)(H)_2] - [C-(C_B)(H)_3] = 52.1 + 23.0 - (-10.2) = 85.3$ kcal/g-mol.

For comparison purposes we examine the aliphatic analog,

$CH_3CH_3 \rightarrow CH_3CH_2^{\cdot} + H^{\bullet}$

for which

$\Delta H = [H^{\bullet}] + [C^{\cdot}-(C)(H)_2] - [C-(C)(H)_3] = 98.1$

The difference between the two energies

$[C^{\cdot}-(C)(H)_2] - [C^{\cdot}-(C_B)(H)_2] = 12.8$ kcal/g-mol

is due to the interaction of the free electron with the π-bonding orbital of the benzene ring and is called the resonance stabilization energy (RSE). The RSE is responsible for the relatively low value of several bond dissociation energies in coal. Table 3.1 lists the RSE of α-radicals deriving from larger rings as estimated by Stein et al.(ref. 31).

TABLE 3.1   Resonance stabilization energies (kcal/g-mol) for radical with $CH_2^{\cdot}$. at position A,B,C,D,E (ref. 31).

Compound	A	B	C	D	E
	13.0				
	16.4	14.7			
	18.5	15.3	22.2		
	15.0	15.0	14.0	16.0	16.0

Using the RSE values of Table 3.1 we can estimate the heat of various dissociation reactions as shown in the following examples. For convenience we shall use
the notation Ph_1 = phenyl, Ph_2 = naphthyl, Ph_3 = phenanthryl. Furthermore, Ph_2^2
will denote 2-naphthyl, Ph_3^1 will denote 1-phenanthryl, etc.

Example 3.2

We consider the reaction

$$Ph_2^2-CH_2CH_3 \rightarrow Ph_2^2-CH_2 \cdot + CH_3 \cdot \qquad\qquad (D2')$$

and its aliphatic analog

$$CH_3CH_2CH_3 \rightarrow CH_3CH_2 \cdot + CH_3 \cdot \qquad\qquad (D2a)$$

The heat of (D2a) is well established (ref. 30) as 85, hence using the RSE value
of 14.9, we estimate for reaction (D2'), ΔH = 85-14.9 = 70.1.

Example 3.3

To estimate the heat of reaction

$$Ph_2^2-CH_2-PH_1 \rightarrow Ph_2^2-CH_2 \cdot + Ph_1 \cdot \qquad\qquad (D4')$$

we start from the simpler

$$Ph_1-CH_2-Ph_1 \rightarrow Ph_1-CH_2 \cdot + Ph_1 \cdot \qquad\qquad (D4'')$$

for which ΔH can be computed by group additivity using the values of ref. 30:

$\Delta H(D4'') = [Ph_1-CH_2 \cdot] + [Ph_1 \cdot] - 10[C_B-(C_B)] - 2[C_B-(C)] - [C-(C_B)_2] = 45 + 78.5$
$- 10 \times 3.3 - 2 \times 5.51 - (-4.86) = 84.3$.

The difference between the heats of reactions (D4') and (D4'') is equal to the
difference of the corresponding RSE which is 3.6, therefore $\Delta H(D4')$ = 84.3 - 3.6
= 80.7.

Because of this large value of ΔH, the rate of direct dissociation of methylene
bridges is negligible relative to other pyrolysis reactions. However, two indirect
dissociation mechanisms are much more energetically favorable. The first proceeds
by the addition of hydrogen atoms and other small radicals to the aromatic ring.
The second is operative in the presence of phenolic hydroxyl groups in the ortho
and para position. These two mechanisms will be discussed in following subsections.

Example 3.4

Here we consider reaction (D5) and its aliphatic analog,

$$Ph_3^1-CH_2CH_2-Ph_2^2 \rightarrow Ph_3^1-CH_2 \cdot + Ph_2^2-CH_2 \cdot \qquad\qquad (D5')$$
$$CH_3CH_2CH_2CH_3 \rightarrow CH_3CH_2 \cdot + CH_3CH_2 \cdot \qquad\qquad (D5a)$$

The heat of (D5a) is 82, so that using the appropriate RSE values from Table 3.1
we obtain $\Delta H(D5')$ = 82 - 15.6 - 16.6 = 48.9.

Example 3.5

We next estimate the heats of reactions (D6)-(D8) which involve the dissociation
of carbon-oxygen bonds. Because the heats of formation of the phenoxy radicals
(Ph-O·) are not available we will assume that their resonance stabilization energies
are the same as those of the corresponding benzyl radicals,

$$[O \cdot -(C)] - [O \cdot -(C_B)] = [C \cdot -(C)(H)_2] - [C \cdot -(C_B)(H)_2] = 12.8$$

The value of $[O \cdot -(C)]$ can be calculated from the heat of formation of $CH_3CH_2O \cdot$,
$$\Delta_fH[CH_3CH_2O \cdot] = [O \cdot -(C)] + [C-(C)(H)_2(O \cdot)] + [C-(C)(H)_3]$$
In this equation $\Delta_fH[CH_3CH_2O \cdot]$, $[C-(C)(H)_3]$ are listed in reference 30, while
$[C-(C)(H)_2(O \cdot)]$ may be approximated by $[C-(C)(H)_2(O)]$. The result is $[O \cdot -(C)]$
= 14.3 so that $[O \cdot -(C_B)] = 14.3-12.8 = 1.5$. Using this value in group additivity
calculations we find the heats of reactions (D6),(D7),(D8) to be 68.3, 85.5, 55.6
which vary by no more than +3 kcal from the heats of reactions (D2),(D3), and
(D4), respectively.

Since the aromatic nuclei in coal are substituted to a considerable degree we
must consider the effect of substituents on the heats of the bond dissociation
reactions. Barton and Stein (ref. 32) conducted some model compound studies showing that the heat of reaction (D2) is reduced by about 1 to 2 kcal/g-mol by an
ortho-methyl substituent. Meta and para substituents were found to have little
or no effect. Earlier studies summarized in reference 32 had shown that meta substitution by a methyl group has no effect, para has a slight effect and ortho has
a more significant effect, up to 3 kcal/g-mol. An investigation of the pyrolysis
of cresols (ref. 33) showed that o-cresol reacts faster than m-cresol by a
factor of about three at 806°K. In both cases the rate determining step was the
dissociation of a benzylic hydrogen. Assuming equal A-factors, the activation
energy in the ortho-compound must be lower than that of the meta-compound by about
2 kcal/g-mol.

All the above calculations refer to a temperature of 300°K. To calculate the
heats of reaction at higher temperatures, eg. 800°K, requires the pertinent heat
capacities. These can be computed by group additivity using the data of reference 30. Using the symbol $[\ ]_C$ to denote the heat capacity of a group, we illustrate the calculation for the reaction of example 3.2 above.
$$\Delta c_p(300°K) = [CH_3 \cdot]_C + [C \cdot -(C_B)(H)_2]_C - [C-(C)(H)_3]_C - [C-(C_B)(C)(H)_2]_C =$$
3.26 cal/g-mol °K.
The corresponding values at 400,500,600 and 800°K are 2.07,1.32,0.62,0.03. Using
these values we obtain
$$\Delta H(800°K) - \Delta H(300°K) = \int_{300}^{800} \Delta c_p(T)dt \approx 0.58 \text{ kcal/g-mol}$$
A similar calculation for reaction (D5´) gives $\Delta H(800°K) - \Delta H(300°K) = 0.08$. These
differences turn out to be quite small.

3.1.2 A-factors

The A-factor is given in terms of the entropy of activation by Eq.(3.2). In
gas phase kinetics $\Delta S°^{\dagger}$ can be broken down to contributions from various degrees
of freedom which can be estimated by the techniques of ref. 30 as shown in several
examples below. However, this procedure cannot be rigorously applied to coal pyrolysis because the reaction medium is an amorphous solid or a viscous liquid, therefore the partition function cannot be factored out as in the ideal gas case. For

certain reactions that can take place in solution as well as in the gas phase it has been found experimentally that as long as the solvent does not play any chemical role, the A-factor in solution was comparable to the A-factor in the gas phase (ref. 34). In a recent study (ref. 35) the rate constants of reactions (D5) and (D8) were estimated to be 0.42 and 1.14 in the liquid and gas phases respectively. However, the experimental data base for these estimates was limited. Moreover, it is not known whether the rough equivalence of rate constants also applies to a condensed phase which is an amorphous solid or a very viscous liquid. In the absence of better information it can still be assumed that A-factors computed for gas phase kinetics are roughly applicable to the coal phase at least when the reactions are not diffusion limited.

In carrying out the detailed calculation of the entropy of activation it must be kept in mind that the reacting functional group is actually attached to a rather bulky molecule. For example, in reaction (D2) Ph is an aromatic nucleus such as naphthalene bridged to other similar units. A more descriptive notation for this reaction then would be

$$CM-Ph-CH_2CH_3 \rightarrow CM-Ph-CH_2^\bullet + CH_3^\bullet$$

where Ph is connected by one or more bridges to other sections of the coal molecule denoted here as CM. The reactant coal molecule may have molecular weight from one to five thousand or more in the case of pyrolysis but less in the case of hydropyrolysis and liquefaction. With such large molecular weights the differences between the reactant molecule and the transition state

$$CM-Ph-CH_2 \cdots CH_3$$

are small as far as rotational partition functions are concerned. The external symmetry is likewise the same, one, in both reactant and transition state. As a result of the large size and complex nature of the molecule, the main contributions to the entropy of activation come from vibrations, mainly bending modes, and internal rotations.

A persistent difficulty in estimating A-factors for condensed phase reactions is the possibility of diffusional limitations. This difficulty is especially pronounced in coal pyrolysis because of the high viscosity of the condensed phase. Consider for example reaction (D5) written in the more detailed form

$$Ph-CH_2CH_2-Ph' \underset{k_1'}{\overset{k_1}{\rightleftarrows}} \overline{Ph-CH_2^\bullet \ \ ^\bullet CH_2-Ph'} \underset{k_2}{\overset{k_2'}{\rightleftarrows}} Ph-CH_2^\bullet + Ph'-CH_2^\bullet$$

where the bar indicates a couple of radicals in close proximity. Because of their bulky size and the high viscosity of the coal phase the two radicals may recombine before diffusing sufficiently apart from each other. This "cage" effect can be analyzed by applying the steady state approximation to the intermediate configuration to obtain for the net dissociation rate the expression

$$r = \frac{k_1 k_2}{k_1' + k_2'} [Ph-CH_2-CH_2-Ph'] - \frac{k_1' k_2}{k_1' + k_2'} [Ph-CH_2^\bullet][Ph'-CH_2^\bullet]$$

under diffusion limited conditions $k_1'' \gg k_2''$ and the above expression simplifies to

$r = k_2'' K_1 [Ph-CH_2-CH_2-Ph^{\wedge}] - k_2 [Ph-CH_2^{\bullet}][Ph^{\wedge}-CH_2^{\bullet}]$

where $K_1 = k_1/k_1''$. The diffusion parameters k_1, k_2 have been calculated theoretically for the case of polymerization reactions (refs. 36-38). In the case of pyrolysis such theoretical calculations are not feasible. We will return to this
subject in the chapter dealing with kinetic modeling.

We shall now present examples of estimation of A-factors for several of the
dissociation reactions (D1)-(D8). It should be emphasized once more that the
assignment of frequencies to internal rotation and bending modes is subject to a
great deal of uncertainty. This is the main reason for the fact that theoretically
calculated A-factors can be easily in error by a factor of ten and in some cases
more. If on the other hand it is possible to exploit analogies with experimentally
known A-factors much more reliable estimates can be obtained. The examples below
illustrate the adjustments required in exploiting such analogies. The various
frequencies are assigned using the data of ref. 30.

Example 3.6

In this example we will estimate the A-factor of reaction (D2). The starting
point is the reaction with Ph an unattached phenyl, Ph_1, for which the experimental
value $\log A = 15.3$ at $1,000°K$ is given in ref. 29. The estimation of the entropy
of activation is based on the change between reactant and transition state,

$$Ph_1-CH_2CH_3 \rightarrow Ph_1-CH_2\cdots CH_3 \qquad (d2)$$

where the dots indicate the partially broken bond. The contribution of the various degrees of freedom are as follows:

(i)   Translation, spin and symmetry make no contribution as being the same in
reactant and transition state.

(ii)   Two principal moments of inertia are increased by a factor of 1.8 with a
$R\ln 1.8 = 1.2$ rotational contribution.

(iii) A C-C stretch ($1,000$ cm^{-1}) becomes the reaction coordinate contributing -1.4.

(iv)   The internal rotation about the Ph_1-C bond (barrier changes from 2 to 15
kcal) contributes -1.8; the internal rotation about the CH_2-CH_3 bond (barrier
changes from 3 to 0 kcal) contributes 0.2.

(v)   There are four bending modes about the CH_2-CH_3 bond. The frequencies of
these modes in the reactant can be estimated by analogy to the aliphatic analog
CH_3-CH_3 for which ref. 30 gives $1,000$ cm^{-1}. Since the reduced mass of $Ph-CH_2CH_3$
is about two times larger, the associated frequencies are about 700 cm^{-1}. The
corresponding frequencies in the transition state can be used as adjustable parameters to bring the estimated total $\Delta S°^{\dagger}$ in accord with the experimental value. The
required frequencies turn out to be 210 cm^{-1}.

We now consider the case when Ph is part of a more bulky coal molecule with molecular weight about 500,

$$CM-Ph-CH_2CH_3 \rightarrow CM-Ph-CH_2 \cdots CH_3 \tag{d2´}$$

The only difference with (d2) is that the contribution of rotation is now negligible. The bending frequencies remain the same because the reduced mass changes by the same factor in the reactant and the transition state. At 1000°K, $\Delta S°^+$ = 5.7 - 1.2 = 4.5 and log A = 15.0. Repeating the calculations at 800°K we obtain $\Delta S°^+$ = 6.4, log A = 15.0.

Example 3.7

To estimate the A-factor for reaction (D5) we start with the base case

$$Ph_1-CH_2CH_2-Ph_1 \rightarrow Ph_1-CH_2 \cdots CH_2-Ph_1 \tag{d5}$$

for which ref. 29 gives log A = 14.4 or $\Delta S°^+$ = 3.0 at 1000°K. An analysis of $\Delta S°^+$ can provide as before a crude estimate for the contribution of the bending modes:

(i)   Translation, spin and symmetry do not change.

(ii)   Rotation: two moments increase by a factor of two contributing $R\ell n2$ = 1.4.

(iii) A C-C stretch becomes the reaction coordinate contributing - 1.4 units.

(iv)   Two internal rotations about the Ph_1-C bonds (barrier changes from 2 to 15 kcal) contribute 2 x (-1.8) = -3.6 units; an internal rotation about the CH_2-CH_2 bond becomes a free rotation resulting in an entropy change of 0.2 units.

(v)   To match the experimental $\Delta S°^+$ = 3 an additional 6.4 units are needed, which can be assigned to four bending modes. If the frequency of each of these bonds in the reactant is 400 cm^{-1}, the frequency in the transition state must be about 180 cm^{-1} We now consider the reaction

$$CM-Ph-CH_2CH_2-Ph´-CM \tag{d5´}$$

differing from d_5 mainly in the molecular weight of the group $CM-Ph-CH_2$. If this weight is taken as 500 the bending frequencies are 400 $(91/500)^{½}$ = 170 cm^{-1} 180 $(91/500)^{¼}$ = 77 cm^{-1}. Using these frequencies and recognizing as before that rotation makes a negligible contribution we find that for 800°K, $\Delta S°^+$ = 1.1 and log A = 13.9.

Example 3.8

We now go back to reaction (D4),

$$CM-Ph-CH_2-Ph-CM \rightarrow CM-Ph-CH_2 \cdots Ph´-CM \tag{d4}$$

where we distinguish the following contributions at 800°K,

(i)   Translation, spin and symmetry do not change.

(ii)   The rotational change is negligible.

(iii) The internal rotations about the $Ph-CH_2$ bond and the $CH_2-Ph´$ bond contribute -2.1 and 0.2 units, respectively.

(iv)   The four bending modes will be taken as in the previous example 170 cm^{-1} in the reactant and 77 cm^{-1} in the transition state making a contribution of 4 x 1.5

= 6 units. The sum of these components is $\Delta S^{\circ\dagger}$ = 3.0 and log A = 14.3.

We have analyzed, rather crudely, the reactions (D2), (D4) and (D5). The A-factor for (D1) is estimated by adjusting the experimental value of log A = 15.5 at 1000°K reported for the dissociation of toluene. The A-factors of (D6)-(D8) can be assumed to be approximately equal to those of (D2)-(D5) respectively. The results of these estimates are all listed in the last table of the chapter.

3.1.3 The effect of phenolic hydroxyl groups

It was mentioned earlier that pehnolic hydroxyl groups have a profound effect on the rates of dissociation reactions. The reactions under consideration are

$HO-Ph-CH_2-X \rightarrow HO-Ph-CH_2\cdot + X\cdot$

where the -OH is ortho or para to the benzylic carbon and X is one of the groups H, CH_3, Ph, CH_2-Ph. In an early study (ref. 33) of cresol pyrolysis, o-cresol and p-cresol were found to decompose about four times faster than m-cresol at 816°K. The rate determining step in each case was the dissociation of a benzylic H atom. Assuming equal A-factors, the activation energy of the first two reactions must have been about 3 kcal lower than that of the last reaction. Although not examined in this study, the dissociation of the benzylic hydrogen in toluene and m-cresol must have very similar activation energies.

The activating mechanism of ortho or para situated hydroxyl was recently identified as due to a keto-enol tautomerism (ref. 39),

Assuming the second step to be rate determining, the effective reaction rate constant is k_2k_1/k_1'. The activation energy of step 2 was estimated as about only 45 kcal/g-mol while the equilibrium constant k_1/k_1' was estimated as 10^{-6} at 400°C. Assuming an A-factor equal to that of bibenzyl dissociation, the effective rate constant k_2k_1/k_1' turns out to be several orders of magnitude larger than that of direct dissociation.

3.2 DISSOCIATION OF FREE RADICALS

The following are representative of this class of reactions:

$Ph-\dot{C}HCH_3 \rightarrow Ph-CH = CH_2 + H\cdot$ (DB1)

$Ph-\dot{C}HCH_2CH_3 \rightarrow Ph-CH = CH_2 + CH_3\cdot$ (DB2)

$Ph-CH_2CH_2\dot{C}H_2 \rightarrow Ph-CH_2\cdot + CH_2 = CH_2$ (DB3)

$Ph-\dot{C}HCH_2-Ph´ \rightarrow Ph-CH = CH-Ph´ + H\cdot$ (DB4)

Among the above reactions (DB1), (DB2) and (DB4) involve the conversion of an alpha radical to a higher energy radical. The attendant loss of resonance stabilization energy results in a higher activation energy compared to the corresponding aliphatic analog. For example reaction (DB2) has an activation energy of 45 compared to 34 for the aliphatic analog

$CH_3CH_2\dot{C}HCH_3 \rightarrow CH_3\cdot + CH_2 = CHCH_3$

The exception is reaction (DB3) where resonance stabilization energy is gained and the activation energy is very low (9.6). The examples below focus on the estimation of A-factors, using the same techniques as in section 3.1.2. The estimation of activation energies can be carried out in a straightforward way by group additivity and is illustrated only in example 3.9.

Example 3.9

We start with reaction (DB1) and compare

$CM-Ph-\dot{C}HCH_3 \rightarrow CM-Ph-CH\text{-----}CH_2 \cdots H$ (db1)

with the aliphatic analog

$CH_3-\dot{C}HCH_3 \rightarrow CH_3-CH\text{-----}CH_2 \cdots H$ (db1´)

for which the experimental value log A = 14.3 at 500°K is reported (ref. 29). The two reactions differ mainly in the internal rotation, Ph-CH bond vs. CH_3-CH bond. Taking this difference into account and adjusting for the temperature yields log A = 15.1 at 800°K.

Example 3.10

We next consider reaction (DB2). The heat of reaction can be immediately calculated by group additivity using the group values of ref. 30 as $\Delta H = 37.8$. The activation energy is given by $E = \Delta H + E´$ where $E´$ is that of the reverse reaction. The latter is about 7.2, section 3.5, therefore E = 45.

To estimate the A-factor we consider the entropy changes associated with

$CM-Ph-\dot{C}HCH_2CH_3 \rightarrow CM-Ph-CH\text{-----}CH_2 \cdots CH_3$ (db2)

and the aliphatic analog

$CH_3\dot{C}HCH_2CH_3 \rightarrow CH_3CH\text{-----}CH_2 \cdots CH_3$ (db2´)

for which ref. 29 gives the experimental value log A = 14.2 at 570°K. The difference in the entropies of (db2) and (db2´) is mainly due to (i) rotation which contributes to (db2´) but not (db2) and (ii) internal rotation in (db2) is about the Ph-$\dot{C}$ bond while in (db2´) is about the CH_3-$\dot{C}$ bond. The differences due to (i) and (ii) nearly cancel each other so that for (db2), log A=14.2 at 570°K and log A=14.4 at 800°K.

Example 3.11

To estimate the A-factor of reaction (DB3) a comparison is made between

$$CM-Ph-CH_2CH_2CH_2\cdot \rightarrow CM-Ph-CH_2\cdots CH_2\cdots CH_2 \qquad \text{(db3)}$$

$$CH_3CH_2CH_2\overset{\cdot}{C}H_2 \rightarrow CH_3CH_2\cdots CH_2\cdots CH_2 \qquad \text{(db3´)}$$

The predominant difference between the two activation entropies are due to (i) rotation - which is negligible in (db3) - and (ii) bending modes about the weakened CH_2-CH_2 bond. In the absence of any information it is assumed that the contribution of the bending modes to $\Delta S\circ^{\dagger}$ is the same in (db3) and (db3´). The difference due to rotation is about 1.6 units, hence we set log A = 14.0 at 650°K or 14.2 at 800°K.

Example 3.12

The last reaction (DB4) involves the entropy of

$$CM-Ph-\overset{\cdot}{C}HCH_2-Ph´-CM \rightarrow CM-Ph-CH\cdots CH-Ph´-CM$$
$$H\cdot$$

consisting of the following parts:

(i)   One C-H stretch becomes the reaction coordinate with a -0.1 units change.

(ii)   One C-C stretch becomes a C$\cdots$C stretch with a -0.5 change.

(iii) Two C-C-H bends become C-C$\cdots$H bends with 2 x 0.4 = 0.8 change.

(iv)   The barrier to internal rotation about the Ph-C bond changes from 15 to 2 kcal contributing 2.1 units. The internal rotation about the $\overset{\cdot}{C}$-C bond becomes a torsion about the $\cdots$ bond contributing -6.1 units.

The total change at 800°K is $\Delta S\circ^{\dagger}$ = -3.8 or log A = 12.8.

3.3   RECOMBINATION OF ALPHA RADICALS

This is the reverse of reaction (D5),

$$Ph-CH_2\cdot + Ph´-CH_2\cdot \rightarrow Ph-CH_2CH_2Ph´ \qquad \text{(R1)}$$

As mentioned earlier, reactions (D5) and (R1) are diffusion limited when they take place in the condensed coal phase. The A-factor estimated below assumes diffusion free recombination, therefore is strictly limited to gas phase recombination.

Under gas phase conditions the activation energy of (R1) is zero. The activation entropy is related to the activation entropy of the reverse reaction as follows

$$\Delta S\circ^{\dagger}(R1) = \Delta S\circ^{\dagger}(D5) + \Delta S\circ(R1)$$

where $\Delta S\circ(R1)$ is the entropy of the reaction. $\Delta S\circ(R1)$ can be estimated by group additivity. At 800°K we have

$$\Delta S\circ(R1) = 2[C-(C_B)(C)(H)_2]_e - 2[C\cdot-(C_B)(H)_2]_e = -35.3$$

where the subscript e denotes entropy. We also have from section 3.1, $\Delta S\circ^{\dagger}(D5) = 1.1$ so that $\Delta S\circ^{\dagger}(R1) = 1.1 - 35.3 = -34.2$. For a bimolecular reaction

$$A = \frac{k_BT}{h} \frac{RT}{P_0} e^2 \exp(\Delta S\circ^{\dagger}/R)$$

with the standard state being at pressure $p_0 = 1$ at. For the entropy change of -34.2, log A = 8.4 ℓ/s g-mol.

3.4 HYDROGEN ABSTRACTION

The following are a few representative reactions

$$X\cdot + Ph\text{-}CH_3 \rightarrow XH + Ph\text{-}CH_2\cdot$$

X = H	(HA1)
X = CH_3	(HA2)
X = C_2H_5	(HA3)

$$X\cdot + Ph\text{-}CH_2CH_3 \rightarrow XH + Ph\text{-}CH_2CH_2\cdot$$

X = H	(HA4)
X = CH_3	(HA5)
X = C_2H_5	(HA6)

$$Ph\text{-}CH_2CH_2\cdot + Ph'\text{-}CH_3 \rightarrow Ph\text{-}CH_2CH_3 + Ph'\text{-}CH_2\cdot \qquad (HA7)$$

The activation energies of (HA1), (HA2) have been estimated as 2.3, 8 by a variant of the bond-order-bond-energy method (ref. 40). The activation energy of (HA3) can be estimated by analogy to suitable aliphatic analogs as follows. The two reactions

$$CH_3\cdot + (CH_3)_3CH \rightarrow CH_4 + (CH_3)_3C\cdot$$
$$C_2H_5\cdot + (CH_3)_3CH \rightarrow C_2H_6 + (CH_3)_3C\cdot$$

have activation energies 8 and 8.9 respectively. The difference of 0.9 is assumed to apply to the reactions (HA2) and (HA2) as well, producing the estimate 8.9 for (HA2).

The activation energies of reactions (HA4)-(HA6) are taken equal to those of their aliphatic analogs. For example, the aliphatic analog of (HA5) is

$$CH_3\cdot + CH_3CH_2CH_3 \rightarrow CH_3CH_2CH_2\cdot + CH_4$$

Finally, the activation energy of (HA7) is taken equal to that of (HA3).

The estimation of A-factors for hydrogen abstraction reactions is illustrated by the following two examples.

Example 3.13

The transition relevant to (HA1) is

$$CM\text{-}Ph\text{-}CH_3 + H\cdot \rightarrow CM\text{-}Ph\text{-}CH_2\cdot \cdots H \cdots H \qquad (ha1)$$

By definition, the entropy of this transition is

$$\Delta S\circ^\dagger = S\circ^\dagger - S\circ(R) - S\circ(H\cdot)$$

where R represents $CM\text{-}Ph\text{-}CH_3$. The difference $S\circ^\dagger - S\circ(R)$ has the following components:

(i)   Translation, external symmetry and rotation zero or negligible.

(ii) Spin from 2 to 1 with contribution Rℓn2.

(iii) The barrier for internal rotation about the Ph-C bond changes from 0 to 7.5 kcal contributing -1.3 units. At the same time a loss of a symmetry factor of 3 associated with internal rotation contributes $R\ln 3$ units.

(iv) A C-H stretch becomes reaction coordinate causing an entropy change of -0.1. A new H$\cdots$H stretch (2800 cm^{-1}) contributes 0.1.

(v)   Two new H$\cdots$H$\cdots$C bends (1000 cm^{-1}) contribute $2 \times 1.1 = 2.2$.

(vi) Two H-C-H Bends (1400 cm^{-1}) become H$\cdots$C-H bends (1000 cm^{-1}) causing an entropy change of $2 \times 0.5 = 1$.

The resultant of these components is $S°^{\dagger}-S°(R) = 5.5$. At the same time standard tables (ref. 30) give $S°(H\cdot) = 32.3$ so that $\Delta S°^{\dagger} = -26.8$ and log A = 10.0.

The A-factors of reactions HA2 and HA3 are best estimated by comparison with aliphatic analog reactions

$$H\cdot + C_2H_6 \rightarrow H_2 + C_2H_5\cdot \qquad \text{(HA1´)}$$
$$CH_3\cdot + C_2H_6 \rightarrow CH_4 + C_2H_5\cdot \qquad \text{(HA2´)}$$
$$C_2H_5\cdot + C_2H_5COC_2H_5 \rightarrow C_2H_6 + C_2H_5COC_2H_4\cdot \qquad \text{(HA3´)}$$

The experimental log A for these reactions are 11.0, 8.5, 8.0 at 400°K (ref. 30). Assuming that log A (HAi)-log A (HA1)=log A (HAi´)-log A (HA1´) for i = 1,2,3 and that each difference is independent of temperature we obtain for 800°K the crude estimates log A (HA2) = 7.5, log A (HA3) = 7.0.

Example 3.14

For reaction (HA4) we need the entropy change of

$$H\cdot + CM\text{-}Ph\text{-}CH_2CH_3 \rightarrow CM\text{-}Ph\text{-}CH_2CH_2\cdots H\cdots H \qquad \text{(ha4)}$$

This change can be calculated by analogy to (ha1) from which it differs only by the fact that the barrier for internal rotation about the Ph-C bond does not change. Therefore, we set $\Delta S°^{\dagger}(HA4) = \Delta S°^{\dagger}(HA1) + 1.3 = 25.5$ and log A (HA4) = 10.3.

Reactions (HA5) and (HA6) can be treated similarly to (HA2) and (HA3). The result is log A (HA5) = 7.8, log A (HA6) = 7.3. Finally, in the absence of any other information we can crudely set log A (HA7) = log A (HA6).

3.5   ADDITION OF RADICALS TO DOUBLE BONDS

These reactions are the reverse of (DB1)-(DB4) and may be exemplified by:

$$X\cdot + Ph\text{-}CH = CH_2 \rightarrow Ph\text{-}\overset{\cdot}{C}HCH_2X \qquad \text{(AD1)}$$
$$X\cdot + Ph\text{-}CH = CH\text{-}Ph´ \rightarrow Ph\text{-}\overset{\cdot}{C}HCH\text{-}Ph´ \qquad \text{(AD2)}$$
$$\underset{X}{\vert}$$

where X = H,CH$_3$ or C$_2$H$_5$. The activation energies of (A1) and (A2) can be approximated by those of the aliphatic analogs

$$X\cdot + CH_3CH_2CH = CH_2 \rightarrow CH_3CH_2\overset{\cdot}{C}HCH_2X \qquad \text{(AD1´)}$$
$$X\cdot + CH_3CH = CHCH_3 \rightarrow CH_3\overset{\cdot}{C}HCHCH_3 \qquad \text{(AD2´)}$$
$$\underset{X}{\vert}$$

These are listed in the compendium of Kerr and Parsonage (ref. 41) as 1.2,7.2, 7.3 for (AD1´) with X = H,CH$_3$, C$_2$H$_5$, respectively. The corresponding values for (AD2´) are listed as 4.3,7.5,8.5.

The entropies of activation of (AD1) and (AD2) can be calculated from the entropy of reaction and the activation entropy of the reverse reaction. Taking as an example (AD1) with X = CH_3 we have

$\Delta S \circ^{\dagger}(AD1) = \Delta S \circ^{\dagger}(DB2) + \Delta S \circ (AD1)$

Using group additivity we calculate for 800°K $\Delta S \circ = -33.7$ so that $\Delta S \circ^{\dagger}(AD1) = 3.4 - 33.7 = -30.3$, log A = 9.3.

3.6   ADDITION OF RADICALS TO AROMATIC RINGS

The limited literature on the addition of small radicals like $CH_3 \cdot$ and $C_2H_5 \cdot$ to aromatic rings has been reviewed by Szwarc et al. (refs. 42,43) and Williams (ref. 44). The reaction mechanism consists of addition to the aromatic ring to form a cyclohexadiene intermediate followed by dissociation as shown in the following two examples:

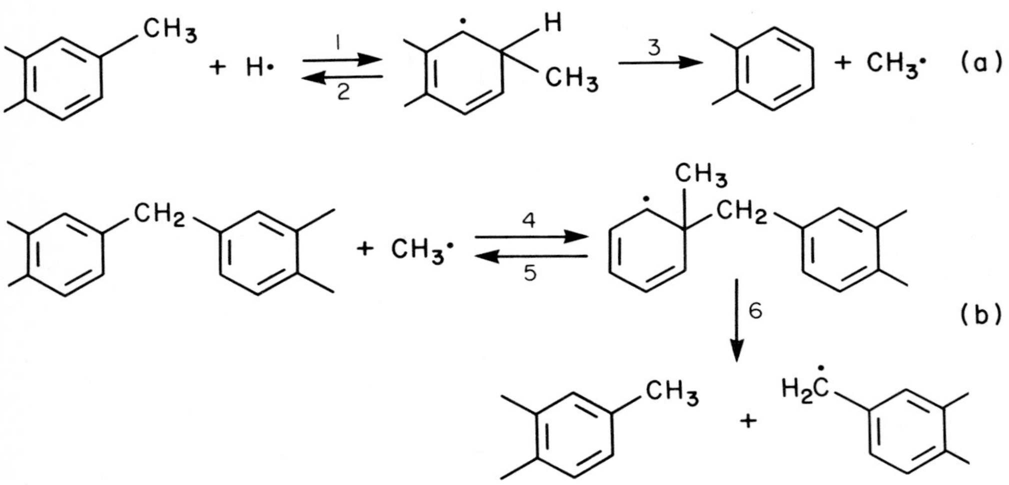

(a)

(b)

Addition reactions have been studied in solution wherein methyl and other carbon-centered radicals were produced by the decomposition of initiators like acetyl peroxide. Hydrogen addition has not been studied as such because it cannot be produced in solution in any clean-cut fashion. However, hydrogen addition at temperatures higher than 400°C has been identified as an important step in the thermal reactions of various aromatic compounds (ref. 33,45).

When addition takes place in solution at modest temperatures, the intermediate cyclodiene radical can recombine with other radicals or abstract hydrogen leading to a variety of products. At temperatures representative of coal pyrolysis (above 400°C) the decomposition steps 2 or 3 in (a) or 5,6 in (b) would be quite rapid to effectively suppress recombination reactions. Because of the larger energy of H· compared to $CH_3 \cdot$ the decomposition step 3 is much faster than (2) and reaction

(a) is kinetically equivalent to

$$Ph-CH_3 + H\cdot \rightarrow PhH + CH_3\cdot \qquad \text{(a')}$$

Likewise, as a result of the higher stability of the benzyl radical

$$Ph-CH_2-Ph' + CH_3\cdot \rightarrow Ph-CH_3 + Ph'-CH_2\cdot \qquad \text{(b')}$$

Because reactions (a) or (b) often follow the route (a') or (b') they are sometimes characterized by the terms aromatic substitution or displacement reactions.

Returning to reaction (a) we note that the addition of the hydrogen atom can
take place at other positions, for example

This reaction is obviously kinetically insignificant. Inasmuch as reported experimental values for the rate constants of addition reactions refer to addition
at any position around the ring, these constants should be suitably reduced if
addition at a specific site is considered. Unfortunately, the relative addition
rates at positions like (a) and (a'') have not yet been measured.

In coal pyrolysis, reactions of the type (a) can be the source of various hydrocarbon gases via the corresponding radicals. For example the $CH_3\cdot$ radical
produced in (a) will evolve as methane after hydrogen abstraction from some suitable site. Reaction (b) is important as a low activation energy route to the
dissociation of methylene bridges.

The rate parameters in Table 3.2 refer to addition to the benzene ring

where X = H, CH_3, C_2H_5 (i=1,2,3) and Y is H or an alkyl chain or bridge. Depending
on the relative stability of $X\cdot$ and $Y\cdot$, the decomposition of the cyclohexadiene
radical would lead to the product PhX or back to the reactant PhY. At high temperatures the dissociation of the hexadiene radical is relatively rapid so that
the overall reaction is controlled by the rate of the addition step. The activation energies of the addition reactions are in the range 0-8 (ref.30), probably
close to 8 for $CH_3\cdot$ and $C_2H_5\cdot$ and about 2-4 for $H\cdot$. The parameter values listed
in Table 3.2 refer to the benzene ring. It must be noted that the rates of addition
increase with increasing size and decreasing stability of the aromatic ring system.
Szwarc and Binks (ref.43) report relative rates of 1,22,27 and 280 for the addition
of methyl radical to benzene, naphthalene, phenanthrene and anthracene, respect-

ively. The rate constants must obviously be adjusted upwards when applied to
kinetic modeling of coal reactions.

3.7   REACTIONS OF CARBOXYL AND PHENOLIC HYDROXYL GROUPS

Among the principal products of pyrolysis, water and carbon dioxide are generally attributed to the phenolic hydroxyl and carboxyl groups in coal. Brooks
et al. (ref.46) used IR spectroscopy and wet chemical methods to measure the concentration of oxygen groups in brown coals undergoing pyrolysis and suggested
that at temperatures below 300°C water is produced by esterification reactions
presumably of the type

$$Ph-COOH + Ph´-OH \rightarrow Ph-COO-Ph´ + H_2O \qquad\qquad (i)$$

At temperatures between 300 and 450°C they observed a further decrease of phenolic
-OH by reactions of the type

$$Ph-OH + Ph´-CH_3 \rightarrow Ph-CH_2-Ph´ + H_2O \qquad\qquad (ii)$$
$$Ph-OH + Ph´-OH \rightarrow Ph-O-Ph´ + H_2O \qquad\qquad (iii)$$

of which (iii) was singled out as more important. At the same time, ester groups
formed by (i) were found to decompose with the elimination of carbon dioxide,

$$Ph-COO-\phi´ \rightarrow Ph-Ph´ + CO_2 \qquad\qquad (iv)$$

Presumably, free carboxyl groups would also decompose under these conditions.

Reactions (i)-(iii) are significant not only as sources of "chemical water"
but as producing additional linkages among the structural units resulting in a
suppression of tar formation. Since the content of hydroxyl and carboxyl groups
is higher in coals of low rank, the above condensation reactions offer an explanation for the low amount of tar produced in the pyrolysis of subbituminous coals
and lignites. The kinetics of the condensation reactions (i)-(iii) have not been
studied at the temperatures of interest to pyrolysis (above 400°C). Moreover,
the experimental evidence has not as yet been sufficient to distinguish which of
these three reactions is more important.

At temperatures above 600°C the reactions of hydroxyl groups become more complex
leading to a variety of products including carbon monoxide. While all reactions
discussed in sections 3.1-3.6 were free radical in nature, the high temperature reactions involving phenolic hydroxyls are believed to largely proceed by concerted
mechanisms. The remainder of the chapter is a survey of concerted mechanisms
relative to the thermal reactions of coal.

Concerted Reactions

The term "concerted" indicates reactions that involve simultaneous breaking
(and making) of more than one bond. By contrast, the free radical reactions surveyed so far involve the breaking (and making) of one bond at a time. The following examples drawn from model compound studies give an idea of the role of concerted reactions in coal pyrolysis.

Cypres and Betten (refs. 47-49) studied the pyrolysis of phenol, o-cresol and p-cresol labelled at specific positions by ^{14}C and ^{3}H. The reactions were carried out in the temperature range 700-900°C and resulted in a wide variety of products. For example, in the pyrolysis of one mole of o-cresol at 750°C for 2.5 seconds, the major products (in moles) were 0.064 benzene, 0.034 toluene, 0.242 phenol, 0.10 water, 0.139 carbon monoxide, 0.128 methane, 0.078 hydrogen and 0.032 char. When the o-cresol was tritiated on the hydroxyl group, the toluene contained a considerable amount of tritium while the water contained 0.61 atoms of tritium. This distribution of tritium was interpreted as evidence against the free radical mechanism (i) which would produce toluene free of tritium. Instead, the authors proposed the concerted mechanism (ii) which produces tritiated toluene. This explanation is not complete, however, since the elimination of oxygen in the last step of (ii) is not mechanistically satisfactory as written down and, in addition, it does not explain the formation of a considerable amount of tritiated water. Mechanism (i) can partially explain the formation of tritiated toluene and water if H• is replaced by H•* as shown in (i')

The relevant issue in choosing between (ii) and (i) is the relative rates of the two reactions producing hydrogen atoms :

Mechanism (ii) provides a better explanation of the products when (iii) is faster than (iii´) while (i´) becomes more credible when (iii´) is faster. Unfortunately the relative rates of (iii) and (iii´) have not been determined independently. Another possible explanation is hydrogen exchange (scrambling) whereby the tritium atom is spread around the ring preceding dissociation. It appears that the ex- perimental data available are not sufficient to discriminate between the free radical and the concerted mechanisms.

To explain the formation of carbon monoxide the authors of refs. 47-49 suggested two possible schemes, one of which involves a seven-membered ring,

Independently of the validity of this particular mechanism, it is clear that the formation of carbon monoxide requires the disruption of the benzene ring and cannot be explained solely by free radical steps. It should be noted in this connection that carbon monoxide appears in significant amounts only when the pyrolysis temp- erature exceeds 700°C.

In an attempt to demonstrate the importance of pericyclic pathways in the thermal reactions of coal, Virk et al. (ref.50) studied the hydrogenation of anthracene, phenanthrene and a high volatile bituminous coal with various hydrogen donor solvents at 300°C for two hours. One of the reactions studied was

The percentage conversion obtained with different H-donor solvents was as follows:

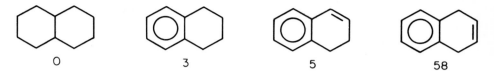

| O | 3 | 5 | 58 |

The large difference with the last two solvents was explained by the Woodward-Hoffman rules governing concerted reactions.

It must be noted that if reaction (iv) were to take place by a free radical mechanism it would require as a first step the dissociation of a hydrogen atom from the H-donor molecule:

$$\text{(structure)} \longrightarrow \text{(structure)} + \text{H}\cdot \qquad \text{(iv')}$$

This reaction has activation energy about 69 kcal, therefore would be extremely slow at 300°C. Using stereospecificity properties, von E. Doering and Rosenthal (ref. 51) demonstrated a concerted path for the thermal decomposition of cis-9, 10-dihydronaphthalene to naphthalene, tetralin and hydrogen.

Several authors who have studied the pyrolysis of tetralin or the dissociation of bibenzyl in tetralin (refs. 52-54) have explained their results solely by free radical mechanisms. More experimental work is needed, preferably with labelled compounds, to determine which of the pyrolysis reactions proceed by concerted mechanisms and which by the more widely accepted free radical mechanisms.

TABLE 3.2   Rate parameters of selected elementary reactions at 800°K

	$\log_{10}A$ (A in s^{-1} or lt/g-mol s)		E (kcal/g-mol)	
D1	14.9	BN	85.3	G
D2	15.3	BN	72.4	G
D3	15.4	BN	69.5	G
D4	14.3	G	80.7	G
D5	13.9	BN	56.4	G
D6	15.3	G	68.6	G
D7	14.3	G	85.5	G
D8	13.9	G	55.6	G

Rxn	$\log_{10}A$		E	
DB1	15.1	G	51.7	G
DB2	14.4	G	45.0	G
DB3	14.2	G	9.6	G
DB4	12.8	G	50.2	G
R1	8.4		0	
HA1	10.0	G	2.3	ZM
HA2	7.5	G	8.0	ZM
HA3	7.0	G	8.9	G
HA4	10.3	G	9.7	BTK
HA5	7.8	G	10.8	BTK
HA6	7.3	G	13.4	ZM
HA7	7.0	G	8.9	G
AD1				
X=1	11.0	G	1.2	G
2	9.3	G	7.2	G
3	9.2	G	7.3	G
AD2				
X=1	--		4.3	G
2	--		7.5	G
3	--		8.5	G
A1				
X=1	9.3	G	1.2	KP
2	7.3	G	7.2	KP
3	7.3	G	7.3	KP

BN: ref. 29
BTK: ref. 30
G: estimated by the author
KP: ref. 41
ZM: ref. 40

Chapter 4

EXPERIMENTAL TECHNIQUES AND RESULTS IN FLASH PYROLYSIS

4.1   EXPERIMENTAL TECHNIQUES

As set out in the general introduction, the survey of experimental results will
be confined to flash pyrolysis at the exclusion of slow pyrolysis or carbonization.
Flash pyrolysis poses three experimental difficulties that need careful consideration: (i) control and measurement of the temperature-time history of the coal particles (ii) suppression of secondary reactions (iii) quantitative collection of
products.

The temperature-time history of the coal particles generally consists of a
heating period, a period at approximately constant temperature and a cooling or
quenching period. While isothermal operation permits the easiest kinetic analysis
of the results, the reaction occurring during the heating and cooling times is often
significant.   The kinetic analysis of the data in this case must take into account
the full temperature-time history. Whether or not the pyrolysis occurs isotherm-
ally, the measurement of the coal particles' temperature is not trivial. In many
cases the temperature cannot be directly measured but must be calculated from a
heat transfer model. The other two experimental problems,the suppression of
secondary reactions and the collection of products, depend on the reactor
geometry and flow pattern and are best discussed separately for the entrained
flow and the captive sample techniques.

4.1.1. The entrained flow technique

In this experimental set-up shown schematically in Fig. 4.1, coal particles
20-100 μm are carried by a primary stream of an inert gas through a water-cooled
injector at the axis of a vertical furnace. A secondary and larger stream of
inert carrier gas flows downward along the furnace under laminar flow to assure
that the particles are not dispersed radially to the furnace walls. Particles
and volatile products are collected by a water-cooled probe of special design
ensuring representative samples. The reaction time is adjusted by the flowrate
of the secondary stream and the axial location of the probe. Heating elements
around the furnace produce a uniform wall temperature in the middle zone of the
furnace. The secondary gas stream is pre-heated to decrease the heating time.

A system of this type has been used by several workers (refs. 55-59). Kobayashi
et al. (ref. 57) discuss in detail the experimental aspects of the technique and
provide an analysis of the heating history of the particles including the effect
of injector geometry and the mixing between primary and secondary gas streams.
Their calculations indicate heating times in the range 5-50 ms. The times required
for 40% weight loss of a bituminous coal varied in the range 10-100 ms as the

furnace temperature ranged from 2,100 to 1,500°K. Since the pyrolysis could not be assumed to occur at constant temperature, the kinetic analysis of the data incorporated suitable heat transfer calculations.

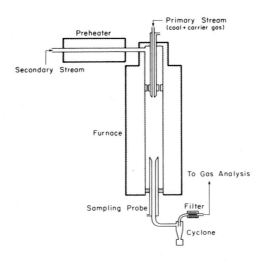

Fig. 4.1. Schematic of an entrained flow pyrolysis furnace. (source: ref. 59).

A second area of concern in the entrained flow technique is the secondary reactions suffered by the volatile reaction products during their residence in the furnace. These reactions make the mechanistic interpretation of product distribution somewhat doubtful but are of no great concern in combustion studies where the primary information required is weight loss and perhaps elemental composition of the volatiles.

The third experimental aspect of the entrained flow technique that requires careful consideration is the collection and analysis of products. In the aforementioned investigation (ref. 57) the weight loss was determined both directly and indirectly. The direct technique consited of simply weighing the char collected in the probe. It was independently shown by cold flow experiments that the collection efficiency of char particles was 95 to 98%, suggesting similar accuracy in the measurements of the weight loss. The indirect technique employed the coal ash as a tracer. The results from this technique are independent of collection efficiency but are subject to some error due to vaporization of mineral matter components.

In addition to the weight loss, certain of the gases (CO, CO_2, SO_2, NO_x) were determined by on-line continuous analyzers (ref. 59) or by gas chromatography (ref. 57). In some studies the solid char was characterized by Fourier-transform infrared spectroscopy (ref. 59). However, products of molecular weight above 200, usually classified as tar, could not be collected quantitatively. Condensation on the probe, the filter, the char, the walls of the gas collection vessel or the walls of the furnace itself preclude a quantitative analysis.

The furnace, injector, probe etc. described in references 57 and 59 represent the most careful and up-to-date designs. Earlier investigations using the entrained flow technique include that of Badzioch and Hawksley (ref. 56) who also used laminar flow conditions and that of Eddinger et al. (ref.55) who used both laminar and turbulent flow conditions. The last study employed a large coal-to-gas mass flow ratio to provide sizeable tar samples. The increased amount of solids, however, interferred with the temperature measurements and increased the extent of secondary reactions.

4.1.2   The captive sample technique

A good description of this technique can be found in Anthony et al. (ref. 60). A small sample (5-200 mg) of ground coal is placed between the folds of a wire-cloth screen (see Fig. 4.2a) heated by a DC or AC current I(t). The resistive assembly is attached inside a metal shell containing an inert gas or hydrogen. The gas pressure can be varied from vacuum to 100 at. In addition to providing the inert atmosphere at the desired pressure, the metal shell serves as a product collection vessel.

The coal sample is placed at the center of the screen where the temperature profile is relatively uniform. Near the electrodes the temperature drops due to conductive losses. A fast response (low mass) thermocouple serves to measure the temperature-time history. The sample size is chosen as a compromise of two considerations. A large sample minimizes the effect of coal inhomogeneity and generates a sufficient quantity of products for analysis. At the same time, however, a large sample of particles creates some irreproducibility because of possible rearrangement of the particles on the screen during heating. Samples of about 200-500 mg have been used by Solomon and Colket (ref.61) and Gavalas and Wilks (ref.62) allowing heating rates as high as 500°C/s and final temperatures as high as 1200°C. In the experiments of Anthony et al. (ref.60) the sample size was 5-10 mg producing heating rates up to 10,000°C/s with final temperatures as high as 1200°C.

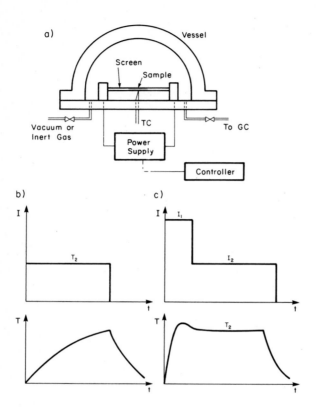

Fig. 4.2. The captive sample technique: (a) pyrolysis apparatus (b) single-pulse T-t response (c) double-pulse T-t response.

The temperature-time history T(t) experienced by the coal sample depends on the current input I(t), the geometry of the resistive assembly and surrounding shell and the nature and pressure of the surrounding gas. For given sample size and gas pressure, and for constant current I, the sample temperature reaches a steady state $T_s = f(I)$ at which the resistive heat input balances losses by conduction, convention and radiation (Fig. 4.2b). The time required to reach the steady temperature is usually on the order of seconds; therefore, constant current inputs are not suitable for producing large heating rates. To circumvent this problem, Anthony et al. (ref.60) applied current inputs consisting of two consecutive pulses of constant current. If I_2 is the current required to produce the steady temperature T_2, then switching from I_1 to I_2 approximately when the temperature has reached T_2, produces the response shown schematically in Fig. 4.2c. The heating rate is thus limited only by the maximum current I_1 of the power supply.

Upon switching from I_1 to I_2, the temperature overshoots slightly before settling to a constant level T_2(Fig. 4.2c). The overshoot can be minimized by a proper choice of the current I_1 and the switching time. After being held at T_2 for the desired time, the sample is cooled by turning the current off. While the cooling period may last for a couple of seconds, after the first 200 ms or so the temperature becomes too low for the pyrolysis reactions to continue.

The most important items in product analysis are weight loss (total volatiles) and tar yield. The weight loss is determined by weighing the coal sample before and after pyrolysis. The tar condenses on aluminum foils lining the walls of the vessel. These foils can be removed and weighed after the completion of an experiment. The tar can be subsequently dissolved in a suitable solvent, e.g. tetrahydrofuran, for elemental analysis or nmr spectroscopy. Because of the short duration of each experiment the pyrolysis vessel remains cold (T <50°C in most cases) and the recovery of tar on the aluminum foils is quite high. A small amount of tar remaining in the gas volume as vapors or aerosol can be collected by flowing the vessel contents through a filter (refs.63,64).

The gaseous products can be analyzed directly from the vessel or after collecting in suitable cold traps. Direct collection was used with sample sizes of 100-200 mg producing measurable concentrations of gases in the vessel (refs.61,62,65). Some experiments (ref.65) indicated that a nonnegligible fraction of the hydro-carbon gases were retained in solution in the tar deposited on the aluminum foil or on various surfaces in the vessel. To recover the gases quantitatively, it was found necessary after each pyrolysis run to heat the vessel to about 150-200°C and thus drive the hydrocarbon gases to the gas phase. The heating in this case was accomplished by a mobile oven sliding on rails (ref.65). A disadvantage of this technique is that at 150-200°C a certain fraction of light products in the tar would also enter the gaseous phase but would not be detected by the gas chroma-tographic procedure employed. The indirect technique of gas collection was de-veloped by Suuberg (refs. 63,64) to concentrate the products derived from small sample sizes (about 10 mg). The contents of the vessel were flown through two cold traps packed with Porapak Q. The first trap maintained at room temperature retained intermediate products such as benzene, toluene and xylene. The second, maintained at -196°C by liquid nitrogen retained all gases except H_2. The products were recovered by warming the traps at 240 and 100°C respectively.

In both the direct and the indirect technique the gaseous products were anal-yzed by gas chromatography employing a thermal conductivity detector for CO, CO_2,

H_2, H_2O and a flame ionization detector for hydrocarbons. The determination of water caused persistent difficulties. One difficulty was the uncontrolled adsorption-desorption of water from pyrolysis or from the atmosphere on vessel walls and gas lines. The other difficulty involved the integration of the broad and distorted chromatographic peak. Solomon and Colket (ref. 61) used a different and somewhat more reliable procedure for water determination.

The overall efficiency of product collection has varied among the various experimental setups employed. The best results seem to have been obtained by Suuberg, who reported (refs. 63,64) a 95% total mass balance and 90% balance on carbon and hydrogen. The largest analysis error is in water, as much as 40%, the lowest in hydrocarbon gases, 5-10%.

An interesting version of the captive sample technique has been described by Jüntgen and van Heek (ref. 66). The main features of the experimental setup was direct connection of the pyrolysis vessel with a time-of-flight mass spectrometer permitting direct on-line analysis of gaseous pyrolysis products and a facility for movie-camera recording of physical changes of the coal particles. A controlled power supply provided a linear temperature-time profile with heating rates up to 3,000°C/s.

4.1.3 The "pyroprobe"

In a variation of the captive sample technique, the coal sample is placed in a heated probe, the "pyroprobe", directly connected to the injection port of a gas chromatograph. The commercially available pyroprobe consists of a platinum ribbon as the heating element with associated power supply and control circuitry. Using coal samples of less than 5 mg, the heating element can supply heating rates up to 20,000°C/s and final temperatures up to 2,000°C. A helium or other inert carrier sweeps the pyrolysis products through a short line into the chromatographic column.

Among the reaction products, gases and compounds of intermediate volatility can be analyzed by using suitable separation columns. For example, capillary columns permit the elution of compounds as heavy as naphthalene and phenol. Heavy tars condense on tube walls or column packing and cannot be analyzed. The char residue is weighed at the conclusion of each experiment. Applications of the pyroprobe have been reported in refs. 67 and 68.

To conclude this section we summarize the relative merits of the three techniques described. The entrained flow technique is suitable for high temperature and short residence time pyrolysis where it provides the best temperature control and rapid quenching. The steady state operation allows the processing of a large quantity of coal to smooth sample inhomogeneity. Although gas and char collection is straightforward, tar collection is difficult. The captive sample technique involves much simpler apparatus and allows arbitrary pressure (including vacuum) and residence time. Except at the highest temperatures, it allows good control of temperature and heating rates. However, high heating rates can be achieved only with small samples (~ 10 mg) exacerbating the problem of sample inhomogeneity. Product collection is good, although the milligram quantity tar collected in the case of small coal samples is insufficient for chemical characterization. Compared to the standard version of the captive sample technique, the pyroprobe arrangement is limited by small sample size and operation at close to atmospheric pressure. In addition, the pyroprobe does not allow the collection of heavy products. However, the direct injection of products into the chromatographic column is a very convenient feature and greatly reduces the turnaround time for an experiment. A more detailed discussion of the operating characteristics of various experimental arrangements is given in a recent comprehensive report by Howard et al. (ref. 69).

4.2 EXPERIMENTAL RESULTS AND DISCUSSION

This section contains a survey of data on weight loss, product distribution and product composition as functions of temperature and pyrolysis time. The bulk of the data reported here derive from essentially isothermal experiments, small particle size and low pressure (vacuum to 1 at). Limited results involving variations in the heating rate and temperature-time history will be discussed in conjunction with the kinetic modeling in chapter 6. Pressure and particle size are variables which affect the rate of transport processes, therefore they will be discussed in chapter 5 on heat and mass transfer. The results discussed in the present chapter relate to conditions which mimimize interferences by transport phenomena and the concommitant secondary reactions. Our survey of experimental data is selective rather than extensive, emphasizing recent comprehensive work and altogether omitting earlier or more narrowly focussed studies. For a broader experimental survey the report of Howard et al. (ref. 69) is highly recommended.

4.2.1 Weight loss

Although only an overall measure of the reaction's progress, the weight loss or total yield of volatiles displays a complex temperature dependence. To facilitate the discussion we introduce a few terms:

cumulative yield: fraction or percentage of the *weight* of a product evolved during the period of pyrolysis relative to the weight of coal on an "as received" or dry ash-free basis; the cumulative yield is generally a function of temperature and time.

instantaneous yield: the derivative with respect to time of the cumulative yield.

yield: the term yield will be used as a short-cut for
                                cumulative yield.

ultimate yield: *at fixed temperature* the cumulative yield asymp-
                                totically reaches a maximum value, *within exper-*
                                mentally reasonable times; this maximum value will
                                be called the ultimate yield and is generally a
                                function of temperature.

   The ultimate yield defined above has relative rather than absolute significance.
Consider for example two first order parallel and independent reactions with the
same product,

$$A \xrightarrow{k_1} P$$

$$B \xrightarrow{k_2} P$$

If A_o, B_o are the initial amounts of A,B, the yield is

$$\frac{A_o}{W}(1-e^{-k_1 t}) + \frac{B_o}{W}(1-e^{-k_2 t})$$

and the theoretical ultimate yield is $(A_o + B_o)/W$. Suppose, however, that k_2 has
a much higher activation energy than k_1 such that at low temperatures $k_2 \ll k_1$.
Then if the pyrolysis time is limited for practical reasons, the ultimate yield
at low temperatures will be A_o/W. At sufficient high temperatures, however, the
constant k_2 will be significant and the ultimate yield will reach its theoretical
value. Since pyrolysis involves a large number of reaction steps with a wide
range of activation energies, the (apparent) ultimate weight loss is an increasing
function of temperature. In principle, the ultimate weight loss also depends on
the temperature-time history because of the coupled nature of the chemical reactions.
However, the dependence on the temperature-time history is rather weak as will be
discussed in Chapter 6.

   Representative weight loss data from an entrained flow system and a captive
sample system are shown in Fig. 4.3 and Figs. 4.4, 4.5 respectively. Figure 4.3
shows the weight loss obtained by the entrained flow technique at high temperatures
and short residence times (ref. 57). The data points represent cumulative weight
loss corresponding to the pyrolysis time under question. Each point, therefore,
corresponds to a distinct experiment employing a different sample of coal particles.
The scatter in the data, as high as 20%, is due to measurement error as well as
sample variation. Another noteworthy feature is the inflection point at the early
stages of pyrolysis, indicating an initial acceleration of the reaction rate.
Since the data have been corrected for the effect of heating time, this initial
acceleration which has also been observed in other studies seems to be related
either to the presence of consecutive reactions or to intraparticle mass transfer retardation.

To prevent secondary reactions of tar vapors on particle surfaces influencing
the weight loss, it is necessary to keep   very low particle density in the entrained flow reactor. Secondary reactions on the reactor walls would still take
place but would only affect the product distribution and not the weight loss.

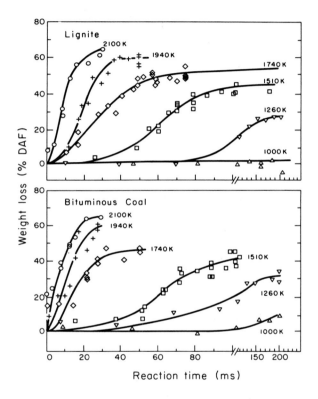

Fig. 4.3. Weight loss vs. pyrolysis time at various furnace
           temperatures (source: ref. 57).

Solomon and collaborators (refs. 61,70) have studied the pyrolysis of a large
number of coals under vacuum (20-60 mm Hg) by the captive sample technique. Figures 4.4,4.5 are representative samples of Solomon's results showing the total
weight loss as a function of pyrolysis temperature for two different residence
times, 20s and 80s. The weight loss (cumulative) has considerable scatter, some
of which may be due to sample variability but a major fraction is probably due
to fluctuations in the sample temperature and loss of fine coal fragments from the
screen. The solid curves were calculated by a kinetic model discussed in Chapter 6.

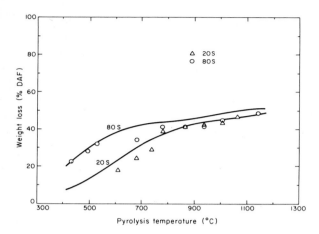

Fig. 4.4. Weight loss vs. temperature for a hvc bituminous
coal "Ohio No. 2" (source: ref. 70).

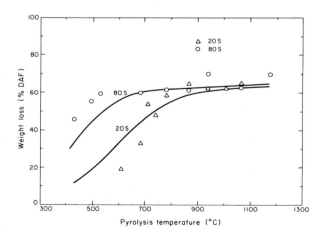

Fig. 4.5. Weight loss vs. temperature for a hva bituminous
coal "Lower Kitanning" (source: ref. 70).

The results of Figure 4.4 exhibit some important trends. Above about 700°C the weight loss at 20 and 80s is the same, within experimental error, indicating that 20s is adequate to attain the ultimate asymptotic value. At temperatures below 700°C the deviation between the results at the two residence times is significant indicating that 20s is insufficient for achieving the ultimate weight loss.

Figure 4.5 presents the weight loss for a bituminous coal with the rank "high volatile A". The main difference with Figure 4.5 is in the ultimate weight loss which seems to change very little with temperature above 900°C. This difference will be explained below in the discussion of individual product yields.

4.2.2   Product distribution

The product distribution is the most essential information relative to the commercial utilization of pyrolysis and at the same time sheds considerable light on reaction mechanisms. Representative experimental data are shown in Figs. 4.6-4.8 taken from Solomon's work (ref. 70). They present the cumulative yields of tar, H_2O, CO_2, CO, H_2 and hydrocarbon gases at residence times of twenty seconds. The balance is the residual solid, char. The label T + M denotes tar and "missing" material that escaped the collection procedure and is estimated only by an overall mass balance. This missing material probably consists largely of tar whence lumped with the collected tar. The solid lines again represent results of model fitting. Despite the cumulative nature of the product yields, the scatter in the data is considerable demonstrating the difficulties inherent in such measurements. Measuring the instantaneous (or differential) yields in this experimental setup is clearly impractical.

The products can be classified into two groups relative to the temperature dependence of the yields. Tar, water and carbon dioxide evolve at lower temperatures with ultimate yields that are essentially independent of temperature above 700°C. The second group of products consisting of gaseous hydrocarbons, carbon monoxide and hydrogen evolve at higher temperatures. The ultimate yield of these products continues increasing with temperature up to 1,000°C or higher.

Coal rank is a very important factor in the distribution and temperature dependence of various products. In bituminous coals, tar makes up 50-80% of the weight loss, the remaining consisting of hydrocarbon gases, water and carbon oxides. In subbituminous coals, water and carbon oxides are produced at increased yields, as much as 60% of the weight loss, while tar contributes only 25-40%. In lignites tar is even lower and gases higher as illustrated in Fig. 4.6.

50

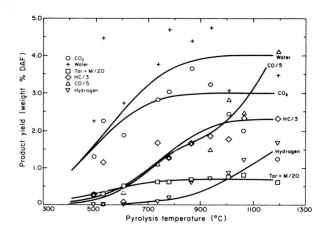

Fig. 4.6. Product yields vs. temperature for a Montana lignite
at 20s pyrolysis time (source: ref. 70).

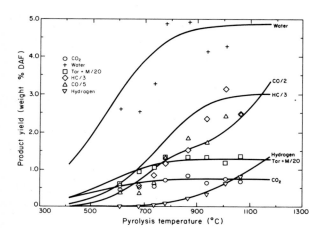

Fig. 4.7. Product yields vs. temperature for a hvc bituminous
coal "Ohio No. 2" at 20s pyrolysis time (source: ref. 70).

The variation of the relative distribution of tar and gases among coals of different rank explains the previously observed temperature dependence of the ultimate yield. In bituminous coals where the weight loss is dominated by tar, the ultimate yield appears to increase very little beyond 700°C. In subbituminous coals and lignites, where a considerable fraction of the volatiles consists of CO and hydrocarbon gases, the ultimate weight loss continues increasing with temperature even beyond 1,000°C.

Tar is the most abundant and commercially valuable product from the pyrolysis of bituminous coals. It is a mixture of many compounds with molecular weights mainly in the range 200-1200. At the temperature of pyrolysis it is produced as a vapor but at room temperature it becomes a viscous liquid or solid. The tar liquids consist largely of dimers of smaller fragments generated by the primary bond dissociation reactions.

In an ideal experimental setup, once released from the coal particles the tar molecules are removed from the high temperature region escaping secondary reactions. The captive sample and entrained flow techniques approximate this desirable operation. Figure 4.9 shows the cumulative tar yield at two pyrolysis times, 20s and 80s, as a function of temperature. Despite the large scatter in the data, the ultimate yield (80s) is clearly independent of temperature above 500°C. As a matter of fact, the yield at 80s seems to slightly decline when the temperature exceeds 800°C. This behavior, which has been observed with several other coals, seems to arise from secondary reactions which are more pronounced at higher temperatures and longer reaction times. It is also observed that the 80s yield exceeds the 20s yield up to about 700°C where they become practically indistinguishable. This indicates that at 700°C (perhaps even lower) a pyrolysis time of 20s is sufficient for the attainment of the ultimate yield of tar.

The temperature-time behavior of the yield of hydrocarbon gases, also shown in Fig. 4.9, is considerably different from that of tar. The ultimate yield increases with temperature in the whole range studied. Moreover, the 80s curve considerably exceeds the 20s curve up to about 900°C indicating the generally higher activation energy of the respective rate determining steps.

Another detailed study of individual product yields was conducted by Suuberg et al., also using the captive sample technique (refs. 63,64,71,72). Because the experiments employed nonisothermal temperature pulses, the measurements do not permit the ready visualization of the dependence on time and temperature, although they are amenable to kinetic analysis as will be discussed in the next chapter. Other results concerning product distribution in flash pyrolysis can be found in refs. 65,67,73-75.

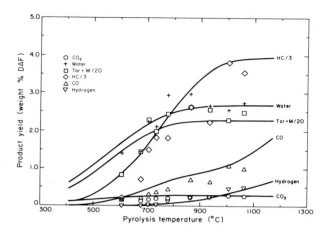

Fig. 4.8. Product yields vs. temperature for a hva bituminous coal "Lower Kittaning" at 20s pyrolysis time (source: ref. 70).

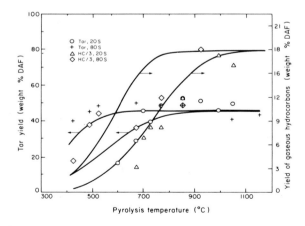

Fig. 4.9. Yields of tar and gaseous hydrocarbons vs. temperature at two pyrolysis times for a hva bituminous coal "Lower Kittaning" (source: ref. 70).

4.2.3. Char and tar composition; distribution of sulfur and nitrogen.

In the previous subsection we presented data on the yields of tar, char and several gaseous compounds. In this subsection we discuss the elemental composition and certain other properties of char and tar with an eye towards their mechanistic significance. We also summarize data about the partition of sulfur and nitrogen among tar, char and gases, which is important to the utilization of these pyrolysis products as fuels.

Figure 4.10 and 4.11 display the elemental composition of char and tar as functions of pyrolysis temperature for a high volatile bituminous coal studied by Solomon (ref. 70). The effect of temperature on hydrogen, oxygen and carbon is quite predictable. Increasing temperature is accompanied by a sharp decrease in the fractions of oxygen and hydrogen due to the evolution of water, carbon oxides and light hydrocarbons, all of which possess O/C or H/C ratios higher than the parent coal. The sulfur content in the char is almost always lower than in the parent coal. However, the temperature dependence of the sulfur varies with the coal examined, probably due to the different amounts of inorganic and organic forms. The nitrogen content in the char is somewhat higher than in the parent coal.

The elemental composition of tar (Fig. 4.11) follows rather different trends. Compared to the parent coal, tar is moderately enriched in hydrogen and sulfur, considerably depleted in oxygen and approximately unchanged in carbon and nitrogen. With increasing temperature, hydrogen decreases slightly, while carbon remains essentially constant. The temperature dependence of oxygen is erratic, perhaps due to the error involved in determination by difference. With some exceptions, the sulfur content is higher than in the parent coal but the temperature dependence is erratic and varies from coal to coal. The content of nitrogen is generally similar to that in the parent coal and shows no noticeable temperature trend.

The distribution of sulfur in the pyrolysis products can be examined in more detail in Fig. 4.12 (ref. 70) which gives organic and total sulfur normalized with respect to the composition in the parent coal. All of the sulfur in the tar is, of course, organic. The normalized organic sulfur is below one in both tar and char, evidently due to the decompositon of reactive sulfur forms like mercaptans and sulfides with the formation of gaseous products, mainly hydrogen sulfide. In some coals with high pyritic content, the organic sulfur content increases with temperature in both the char and tar due to incorporation in the organic structure of decomposing pyritic sulfur. In other coals the organic sulfur shows little temperature trend although it always remains below one. The total sulfur in the char is somewhat less than unity and slightly declines with temperature. References 70 and 76 by Solomon provide detailed results and discussion about the evolution of sulfur forms during pyrolysis.

54

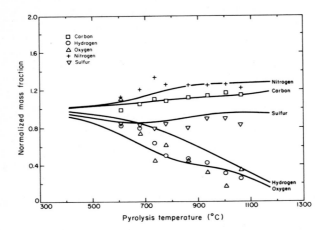

Fig. 4.10. Char composition vs. temperature for a hvc bituminous
coal "Ohio No. 2" at 20s pyrolysis time (source: ref. 70).

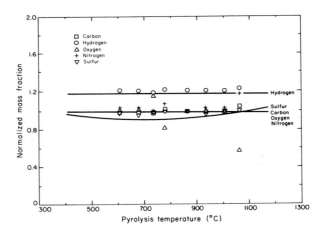

Fig. 4.11. Tar composition vs. temperature for a hvc bituminous coal
"Ohio No. 2" at 20s pyrolysis time (source: ref. 70).

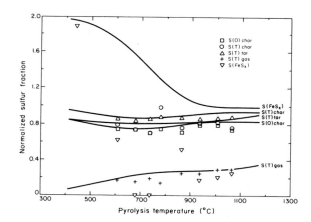

Fig. 4.12. Distribution of sulfur in pyrolysis products of a hvc bituminous coal (Ohio No. 2) vs. temperature at 30s; 0: organic, FeS_x: as sulfide of iron, T: total (source: ref. 70).

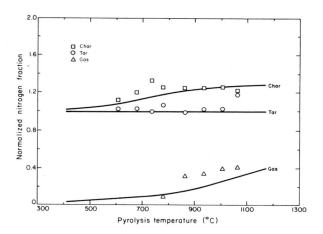

Fig. 4.13. Distribution of nitrogen in pyrolysis products of a hvc bituminous coal (Ohio No. 2) vs. temperature at 30s (source: ref. 70).

Figure 4.13 is an example of the nitrogen distribution in char, tar and gases (ref. 70). The normalized nitrogen in the tar is close to unity indicating that evolving tar molecules contain nitrogen functional groups in the same abundance as in the coal. Char, on the other hand, contains a larger amount of nitrogen while the gases a correspondingly lower amount.

The evolution of nitrogen at higher pyrolysis temperatures typical of pulverized combustion has been studied by Blair et al. (ref. 67) and Pohl and Sarofim (ref. 77). Figure 4.14 shows weight loss and nitrogen loss as a function of temperature, clearly indicating the significant decrease in the nitrogen content of char. Figure 4.15 shows the ultimate weight loss and nitrogen in the char as a function of temperature for coals heated in crucibles. Although these experiments involve slow heating they clearly show the nitrogen content to decline to almost zero as the temperature increases from 1000 to 2000°C.

The interpretation of the results shown in Figs. 14 and 15 is quite straightforward. Below 1000°C nitrogen evolves only with the tar while the gases are almost nitrogen free. As a result the char is gradually enhanced in nitrogen. Above 1000°C various forms of ring nitrogen in the char decompose with the evolution of products such as HCN resulting in a decrease of the nitrogen content.

In addition to the elemental analysis of char and tar, Solomon (ref. 78) and Solomon and Colket (ref. 79) obtained Fourier transform infrared spectra and ^{13}C-^{1}H cross polarization nmr spectra. Figures 4.16 and 4.17 are reproduced from this work. An inspection of Figure 4.16 shows the infrared spectra of coal and tar to be remarkably similar indicating similar types and concentrations of bonds. Figure 4.17 shows the nmr spectra of coal, tar and low temperature (<1000°C) char, in each case the broad peak representing aromatic and the narrow peak aliphatic carbon. Coal and tar have identical values for the aromaticity f_a while char a somewhat higher value. The spectrum for the liquid product, tar, has considerably more detail than the spectra for the solid materials.

Further characterization of the chemical structure of tar was carried out by Gavalas and Oka (ref. 7). Three coals characterized in Table 4.1 were subjected to pyrolysis under vacuum at 500°C for 30s. The tar collected was separated by gel permeation chromatography into three molecular weight fractions: L(>1000), M(700-1000) and S(300-700). Each fraction was characterized by ^{1}H nmr and elemental analysis. The results are shown in Table 4.2 along with corresponding results for tetrahydrofuran extracts of the same coals.

A perusal of Table 4.1 shows coal to be closer in elemental composition to its tar than its extract especially when the CO_2 evolved at 500°C is deducted when computing the elemental composition. Tar has essentially the same carbon content as the coal but somewhat higher hydrogen and slightly lower oxygen contents. The extracts on the other hand have substantially lower oxygen content.

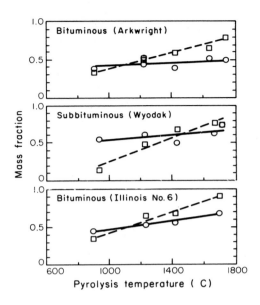

Fig. 4.14. Weight loss (———) and nitrogen loss
(-----) vs. temperature (source: ref. 67).

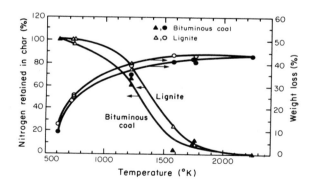

Fig. 4.15. Weight loss and nitrogen retention in char for
coal heated in crucibles (source: ref. 77).

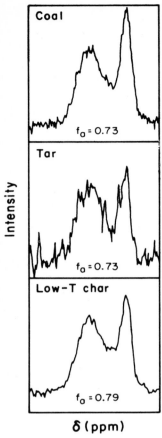

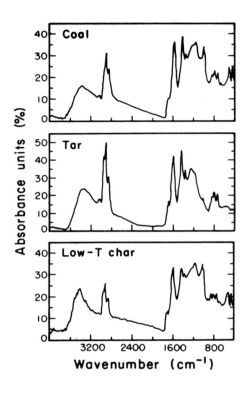

Fig. 4.16. ^{13}C-^1H cross-polarization
spectra of a hva bituminous coal, its
tar and low temperature char (source: ref. 79).

Fig. 4.17. Infrared spectra of a hva
bituminous coal, its tar and low temperature char (source: ref. 79).

The ^1H nmr data of Table 4.1 reveal that more than two-thirds of the hydrogen
is in aliphatic form. However, the split between H_α(alpha) and $H_{\beta+}$(beta or further
from the ring) varies with the rank of the coal. The $H_{\beta+}$ predominates in the
subbituminous coal indicating long chains or hydroaromatic structures while
H_{ar+o}, H_α and $H_{\beta+}$ are present in about the same amounts in the two bituminous
coals.

The molecular weight data of Table 4.2 show the middle fraction (M) to comprise
about half of tar and extract. At the same time, the heavy fraction (H) is more
abundant in the extract while the light fraction (L) is more abundant in the tar.

TABLE 4.1

Elemental composition of coals(daf), extracts and pyrolysis products (source: ref. 7).

		C	H	N	S	O(diff)
Subbituminous C,	c	72.1	5.00	1.28	0.85	20.7
Wyoming Monarch	cc	75.2	5.35	1.36	0.91	17.2
Seam,	e	77.8	8.75	1.28*	0.85*	11.4
PSOC 241	p	74.8	8.0	1.28*	0.85*	15.1
	c	75.7	5.8	1.60	2.89	14.1
HVC Bituminous,	cc	76.2	5.85	1.62	2.93	13.4
Kentucky No.9 Seam	e	82.1	7.4	1.60*	2.89*	6.0
	p	76.1	7.1	1.60*	2.89*	12.3
	c	85.9	5.75	1.48	0.74	6.2
HVB Bituminous,	cc	86.0	5.75	1.49	0.74	6.0
W. Virginia	e	87.5	7.3	1.48*	0.74*	3.1
	p	85.9	7.25	1.48*	0.74*	4.7

c   coal, dry-ash-free; cc   coal after subtracting CO_2 evolved by pyrolysis at 500°C for 30s; e   coal extracts; p   coal pyrolysis products.
*   Assumed for the purpose of estimating the oxygen content.

TABLE 4.2

Elemental analysis, molecular weights and nmr data for extracts and pyrolysates (source: ref. 7).

		Sub-bituminous		HVC Bituminous		HVB Bituminous	
Extract (wt %)[a]		4.5		8.0		1.6	
Pyrolysis (wt %)[b]			6.0		19.1		10.5
GPC* (%)[c]	L	32	26	36	20	32	27
	M	54	48	48	49	44	42
	S	14	26	16	31	24	31
	(%)[d]						
^{1}H nmr	H_α	20	23	32	34	32	37
	$H_{\beta+}$	68	58	40	33	37	31
	H_{ar+0}	12	19	28	33	31	32
	(%)[e]						
^{13}C nmr	C_{ar}	42	–	58	–	–	–
	C_{al}	58	–	42	–	–	–
	(%)						
L	C	76.1	75.6	81.3	75.1	85.7	85.8
	H	8.9	8.1	7.8	7.1	6.9	6.7
M	C	77.5	73.7	81.5	76.5	87.6	85.9
	H	8.7	8.3	7.1	7.1	7.6	8.0
S	C	82.6	76.1	85.5	–	89.5	85.9
	H	8.5	7.3	7.4	–	7.2	6.7

[a] ± 0.5; [b] ± 1.5; [c] ± 1; [d] ± 2; [e] ± 2

* L,M,S: fractions with molecular weights >1000, 700-1000, and 300-700

4.2.4 Effects of pretreatment and atmosphere of pyrolysis

Studies of pretreatment involve exposure to some agent at specified temperature and time, followed by pyrolysis in an inert gas, as usual. By limiting the
temperature and residence time, the extent of thermal decomposition reactions
during pretreatment can be minimized. Howard (ref. 80) has reviewed several
early publications on pretreatment by nitric oxide, steam and oxygen. Pretreatment by nitric oxide at about $300^{\circ}C$ (ref. 81) led to a modest loss of hydrogen,
a modest gain in oxygen and nitrogen and a substantial loss in the proximate
volatile matter. The swelling and agglomerating properties of the coal were
significantly reduced. Nitrous oxide had similar properties while nitrogen
dioxide reacted much faster, even at lower temperatures, and oxidized the coal
more extensively.

Pretreatment by steam has not shown any noteworthy effects. Heating a lignite
at about $300^{\circ}C$ in the presence of steam (ref. 80) produced slight decrease of
weight loss and slight change of product distribution from the subsequent
pyrolysis.

Pretreatment in oxygen has long been studied as a means of reducing the
swelling and agglomerating properties of coal and is, in fact, a necessary step
in some of the gasification processes currently under development. Howard
(ref. 80) has summarized early studies showing that pretreatment by oxygen
increases the yield of carbon oxides and water formed in the subsequent pyrolysis
while it decreases the yield of tar and reduces or eliminates swelling and
agglomeration. The last effect was attributed principally to a change in the
surface of the coal which prevented particle agglomeration even when the particle
interior passed through a plastic phase. Forney et al. (ref. 82) found that
treatment with a gas containing 0.2% oxygen at $400-425^{\circ}C$ for five minutes drasti-
cally reduced swelling and eliminated agglomeration of a caking coal. Under
these conditions, substantial devolatilization could not be avoided. McCarthy
(ref. 83) found that pretreatment in an atmosphere of 2-10% oxygen at $400^{\circ}C$ for
a few seconds similarly greatly reduced agglomeration of a caking bituminous
coal.

A comprehensive study of the effect of preoxidation on subsequent pyrolysis
and the properties of the resulting char was conducted by Mahajan et al. (ref.
84). A strongly caking coal was heated in air at $120-250^{\circ}C$ for a few minutes
to four hours. The weight gain during this treatment increased with temperature
and time, not exceeding 6.5% under the most drastic conditions. Figure 4.18 shows
the weight loss of coal at different levels of prexidation as a function of
pyrolysis temperature. The level of preoxidation was in all cases fully characterized by the percentage weight gain without reference to temperature and duration of the oxidative pretreatment. Below $450^{\circ}C$ the level of preoxination has

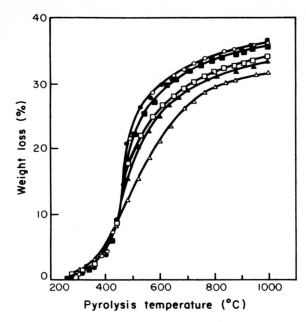

Fig. 4.18. Pyrolysis weight loss vs. preoxidation
level for a caking bituminous coal PSOC 337; weight
gain during preoxidation (%) ● none, o 0.45, ■ 0.75,
□ 1.4, ▲ 2.4, △ 7.0 (source: ref. 84).

little effect on the weight loss of pyrolysis but above 450°C the weight loss
decreases with the preoxidation level. The results shown in Fig. 4.18 are by
no means representative of all caking coals. Working with another caking coal,
the authors found a more complex dependence of the pyrolysis weight loss on the
level of preoxidation. At pyrolysis temperatures less than 500°C, the weight
loss increased with the level of preoxidation but above 500°C the variation
became erratic. The complex dependence of weight loss was attributed to the
following competing factors. Addition of oxygen produces functional groups,
such as carboxyl, which during pyrolysis evolve carbon oxides and water making
a positive contribution to the weight loss. At the same time, the production of
carboxyl and other oxygenated groups reduces the amount of hydrogen that would
otherwise be available for tar generation. To this explanation, one must add that
increased water evolution during pyrolysis increases the degree of crosslinking
leading again to lower tar yield. In addition to the changes in the weight loss,
the chars of preoxidized coals had more open structure, sharply higher CO_2 surface area, and moderately higher reactivity in oxygen.

The pyrolysis experiments discussed until now were all obtained either under
vacuum or under an inert gas such as nitrogen or helium. Pyrolysis in a hydrogen
atmosphere - hydropyrolysis - is important in its own right and will be examined
separately in chapter 7. Pyrolysis in an oxygen atmosphere is an important

63

part of combustion. It is generally believed that oxygen does not affect the primary devolatilization reactions, although it rapidly oxidizes the volatiles in the surrounding gas provided the temperature is sufficiently high as in a combustion furnace. Only subsequent to the rapid volatile release, is oxygen able to reach and oxidize the char particles. The effect of oxygen at lower particle temperatures or when the surrounding gas is cold has not been studied as such.

Pyrolysis in H_2O and CO_2 is of some interest but has received very little attention. Jones et al. (ref. 85) compared the fluidized pyrolysis of a wet (20% moisture) or predried subbituminous coal in nitrogen or steam with the results shown in Table 4.3. The term "liquor" in the table indicates the aquous phase containing the chemical water of pyrolysis along with small amounts of phenolic compounds.

TABLE 4.3

Effect of moisture and fluidizing gas on product yields from the pyrolysis of a subbituminous coal at $1400^{O}F$ in a fluidized bed (ref. 85).

	As received		Dried at $300^{O}F$	
Feed moisture %	20	20	0	0
Fluidized gas	N_2	H_2O	N_2	H_2O
Yields (% dry basis)				
Char	54.6	53.0	56.0	53.1
Tar	11.8	11.0	0.0	9.7
Liquor	7.0	8.7	9.6	11.4
Gas	26.6	27.3	25.4	25.8

The results of Table 4.3 indicate that the tar yield is slightly higher from the moist (as received) coal irrespective of the fluidizing gas. The total weight loss is slightly higher for the moist coal or when steam was the fluidizing gas. Although the differences are small and subject to some scatter, one might venture the following conclusion. Drying coal induces changes in the pore structure or the chemical structure that reduce tar and gas production in subsequent pyrolysis. The observed increase in the yield of liquid (chemical water) when steam is the fluidizing gas is rather puzzling and might be due to errors in the material balances. The nature of the fluidizing gas seems to have little effect on the yields of tar and gases suggesting little effect on secondary tar-cracking reactions. Further discussion on the possible effects of H_2O or CO_2 on secondary reactions will be given in a subsequent subsection dealing with the pyrolysis process of the Occidental Research Corporation.

4.2.5 Effect of inorganic constituents or additives on pyrolysis product yields

Broadly speaking, inorganic matter can operate in two forms to influence the pyrolytic or other reactions of coal. One is as discrete inclusions or continuously distributed material within the coal particles. This form consists of the inherent mineral matter of coal or of material added by impregnation or ion exchange. The second form consists of mechanically mixed inorganic material remaining external to the coal particles. The effects of these two forms of inorganic matter will now be examined separately.

The main groups of minerals in coal include clays (e.g. kaolinite and illite), silica (quartz), sulfides (mainly pyrite), carbonates (e.g. $CaCO_3$), smaller amounts of sulfates and oxides and minor amounts of other minerals. Detailed composition normally refers to the ash, i.e. the inorganic matter remaining after complete oxidation. The transformation of mineral matter to ash involves loss of water from the clays, CO_2 from the carbonates, oxidation of pyrite to iron oxide and fixation of oxides of sulfur in the form of sulfates by calcium and magnesium oxides. The total amount of ash varies widely with coal but generally remains below 25% by weight. The ash composition also varies widely. For example, the composition of the major ash components in American bituminous coals was found in the range SiO_2:20-60%, Al_2O_3:10-35%, Fe_2O_3:5-35%, CaO:1-20%, MgO:0.3-4%, TiO_2:0.5-2.5%, Na_2O + K_2O:1-4% and SO_3:0.1-12% (ref. 86). Subbituminous coals and lignites contain larger amounts of CaO and MgO.

An important consideration relative to the catalytic activity of inherent mineral matter in pyrolysis or gasification is size distribution. Mineral matter is generally present in two forms, either as discrete particles of about one micron size or larger, or distributed on a much finer scale, in association with the organic matter. For example, calcium and magnesium in low rank coals are largely present as cations associated with carboxylic groups. Mahajan (ref. 87) refers to measurements of N_2 surface areas as high as 10^2 m/g for coal minerals corresponding to a mean diameter of about 0.15 μm.

Detailed investigations of the effect of alkaline earth cations on the pyrolysis products of low rank coals have been carried out by Schafer (refs. 88-90). These studies compared acid-demineralized coals with coals that subsequent to demineralization were converted to cation form (Na, K, Mg, Ca, Ba). The measurements included the yields of CO_2, CO and H_2O at different pyrolysis temperatures. In addition, the carboxylic and phenolic content of the coal before and after pyrolysis was determined by titration. The presence of cations was found to alter the relative yields of the three product gases but not to affect the overall weight loss. For example, the ratio H_2O/CO_2 was always smaller for the cation forms compared to the acid form. Significant differences in the gas yields were observed between various cation forms, some possibly due to formation of cyanide compounds by reactions of the cation with the carrier nitrogen. Comparisons between the CO_2, CO evolved and acidic content of the coal led to the conclusion that CO_2 derives from carboxyl groups and CO derives from phenolic groups. Water was presumed to derive from some unidentified oxygen group associated with carboxylic groups. These conclusions are at some variance with the work of Brooks et al. (ref. 46) discussed in section 3.7. In a subsequent investigation of the flash pyrolysis of various low rank coals Tyler and Schafer (ref. 91) found that the presence of cations had profound effect on the yields of tar, C_1-C_3 hydrocarbons and total volatile matter. Removal of cations present in the coals by acid wash increased the yield of tar by as much as a factor of two but had small effect on the yield of hydrocarbon gases. Conversely, addition of calcium ions to the acid-form led to decreased tar and total volatile matter. The fact that the change of the tar yield is not accompanied by a corresponding change of the yield of gases suggests that the effect of cations is not manifested via secondary tar-cracking reactions. Instead it was suggested that the cations might supress tar evolution either by restricting the micropore structure, or by catalyzing the recombination of metaplast molecules before volatilization could take place.

Mahajan and Walker (ref. 92) studied the effect of demineralization by acid treatment on the N_2 and CO_2 surface areas of a number of coals and their carbonization chars. Very divergent trends were observed among the various coals with the surface areas in some cases increasing, in others decreasing and with N_2 and CO_2 areas not necessarily changing in the same direction. Changes in the porous structure can certainly affect tar evolution but the evidence to date is clearly insufficient for firm conclusions. The alternative explanation based on the chemical role of mineral components is certainly plausible, especially considering that the acid treatment of coal to remove cations could also remove or modify the acidity of clay minerals. Such clays might play some role in tar-forming reactions via carbonium ion mechanisms. Further work to delineate the pyrolytic behavior of cation exchanged coals is desirable especially in view of the potential use of cations as catalysts or sulfur scavengers for gasification and combustion.

The effect of inorganic additives in the form of powders mechanically mixed with coal has also been examined. Yeboah et al. (ref. 93) studied the product yields from a bituminous coal and a lignite pyrolyzed in a fluidized bed in the presence of calcined dolomite particles. The presence of the dolomite resulted in decreased tar yield and increased gas yield in all cases. These changes are clearly due to secondary tar cracking reactions on the surface of the dolomite particles.

A different mode of introduction of inorganic additives was investigated by

Franklin et al. (ref. 94). A Pittsburgh No. 8 bituminous coal was demineralized by extraction with a HF-HCl solution and co-slurried in water for for 24 hours with extremely fine particles (0.1 μm) of calcium carbonate or calcium oxide and calcium carbonate. This pretreatment led to the incorporation of about 20% calcium carbonate. While the form of the added material was not determined, it probably consisted of particles coating the surface or penetrating the larger pores of the coal with smaller amounts incorporated on a finer scale by association with the acidic functional groups of coal. The demineralized coal and the calcium-treated coal were subjected to rapid pyrolysis by the captive sample procedure with the following results. Addition of calcium resulted in substantially lower tar yield (20% versus 30%) and lower yield of light hydrocarbon gases, especially at temperatures above $1100^{o}K$. At the same time, calcium treated coals gave considerably higher yields of carbon monoxide, carbon dioxide and water. The overall weight loss in the calcium treated coal was decreased at temperatures above $900^{o}K$. The decreased yield of tar was attributed to secondary reactions of cracking and repolymerization catalyzed by the calcium additive. Such reactions would normally increase the yield of light hydrocarbons. That the yield of these hydrocarbons actually decreased, could be explained by the calcium additive catalyzing the further cracking of methane, ethane, etc. to carbon and hydrogen. Much of the increase in the yield of carbon dioxide was shown to result from the decomposition of calcium carbonate to calcium oxide, a reaction catalyzed by the carbon surface. On the other hand, the increased yield of carbon monoxide was attributed to the decomposition of phenolic groups in the coal structure.

4.2.6 Miscellaneous techniques and results

In this subsection we review some additional pyrolysis studies which did not fit properly in the previous sections. A number of early studies employed thermogravimetric analysis with very low heating rates, a few degrees per minute. The other group of studies employed heating by light, laser light, or plasma arc achieving very high heating rates.

A thermal balance is an apparatus providing continuous measurement of the weight of a static sample under the flow of a carrier gas and linearly rising temperature. Waters (ref. 95) described some early thermal balances and discussed weight loss curves from coal pyrolysis. van Krevelen et al. (ref. 96) used a thermal balance of the torsion type to correlate weight loss with plastic properties. Figure 4.19 shows some typical weight loss curves obtained with a low volatile bituminous coal. The chief features of these curves are (i) The S-shape of the cumulative weight loss. The initial acceleration of the curve, observed also in section 4.2.1, could be due to consecutive reactions or the solubility of some pyrolysis products in the coal melt. (ii) The temperature of maximum devolatilization rate increases with increasing heating rate. More recently, Ciuryla et al. (ref. 97) employed a modern thermobalance to study the weight loss of coals of different ranks under heating rates 40-160°/min. As in earlier studies, maximum rate of devolatilization increased with heating rate. The cumulative yield at a final temperature of 1000°C was found independent of the heating rate, in agreement with the studies reviewed in the previous subsections. The weight loss data were fitted successfully by the Pitt-Anthony model of distributed activation energies.

The principal advantage of the thermogravimetric technique is the continuous and accurate weight measurement. Its chief disadvantages is the inability to operate at high heating rates and constant temperatures and the difficulty to measure accurately the sample temperature. Isothermal operation must always be

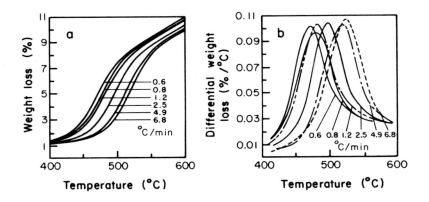

Fig. 4.19. Thermogravimetric analysis of a lo.. volatile bituminous coal. Cumulative (a) and differential (b) weight loss vs. temperature at different heating rates (source: ref. 96).

preceded by a period of slow linear heating. With current commercial models, the sample size can be varied from a few milligrams to about a gram. With large sample size, the sample temperature is uncertain and secondary reactions may become important. Using a small sample size and a sweep gas minimizes secondary reactions but renders product analysis difficult. In fact, none of the aforementioned studies measured individual product yields. Combination of continuous and accurate weight measurement with continuous analysis of gaseous products by infrared and flame ionization detectors seems a most promising technique despite the restriction to low heating rates.

The kinetic feature of constant and slow heating rates has been employed without the gravimetric capability, simply by placing the coal sample in a furnace under the flow of a sweep gas. Using a suitably large sample, of the

order of one gram, it is relatively straightforward to measure the instantaneous
rate of generation of individual gaseous species at the cost, of course, of
allowing secondary reactions on the particle surface. Experiments of this type
have been carried out by Fitzerald and van Krevelen (ref. 98) and Juntgen and
van Heek (ref. 66).

Investigation of pyrolysis in the context of pulverized combustion requires
very high heating rates in the range of 20,000 - 100,000 ^{o}C/s. One experimental
technique for achieving such rates in a small scale apparatus is the irradiation of
coal with ordinary or laser light. Sharkey et al. (ref. 99) irradiated coal
in the form of 3/8" cubes or a fine powder with light from a xenon lamp or a
ruby laser operating at specified power levels. The coal temperature attained
under these conditions could not be determined but was estimated to be in excess
of 1000°C. The product yields obtained under irradiation compared to those
obtained under ordinary carbonization showed much higher contents of acetylene
(absent in carbonization) and carbon oxides and sharply lower content of methane.
These results were attributed to the higher temperatures prevalent under irradiation. However, the yields of hydrocarbons other than methane and acetylene
exhibited irregular and on the whole obscure variations. Evidently, secondary
reactions in the gas phase, as well as in the solid phase, are responsible for
the overall product distribution.

Granger and Ladner (ref. 100) analyzed the gaseous products of several coals
under irradiation from a xenon lamp with and without a filter to remove the UV
component of the light. Their small scale apparatus allowed variation of the
incident energy and their analysis included gaseous products, i.e. carbon oxides
and light hydrocarbons. Water was not determined and tar was estimated rather
crudely. Figure 4.20 shows the major pyrolysis products versus total incident
energy. The principal feature of the results is the rapid increase in hydrogen,
acetylene and carbon monoxide with increasing light intensity. The rapid
increase in carbon monoxide is easily accounted by the fact that this gas is the
only oxygen-containing product after the evolution of carbon dioxide has been
completed in the initial phases of pyrolysis. The latter gas was actually
absent, evidently being converted to carbon dioxide and carbon at the high temperatures of the experiment. The yield of tar was found to pass through a
maximum as a result of secondary cracking reactions. The results of Fig. 4.20
were obtained using UV-filtered radiation. Unfiltered light resulted in decreased
yields of olefins and paraffins and increased yields of carbon monoxide, hydrogen,
acetylene and soot. These changes indicate the presence of photochemical cracking reactions in the gas phase.

A technique for heating coals at very high temperatures and heating rates is
the plasma arc. Bond et al. (ref. 101) used an argon plasma jet to pyrolyse

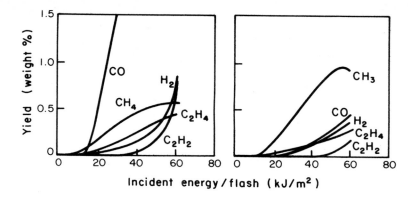

Fig. 4.20. Yields of gaseous products from the flash
pyrolysis by xenon light of a noncaking (a) and
a caking (b) coal (source: ref. 100).

coals of various ranks. The device provided rapid heating and quenching of the
pulverized coal particles. Temperatures at the center of the jet were estimated
to be between 10,000 to 15,000°C. The maximum temperature attained by the coal
particles could not be determined but was estimated to be well in excess of
1000°C. The products were gas and soot with tar completely absent. The gas
contained mainly hydrogen, carbon monoxide and acetylene, which comprised over
95% of all hydrocarbon gases. Acetylene yields as high as 20% by weight of
the coal were obtained. Overall conversion increased with decreasing particle
size and with increasing proximate volatile matter of the starting coal.

A thermodynamic analysis of product distribution from ultrahigh temperature
coal pyrolysis in various atmospheres was made by Griffiths and Standing (ref.
102). Although, in each case the equilibrium mixture contains free radicals as
well as stable species, attention can be limited to stable species assuming
suitable radical recombination during quenching. At temperatures above 1800°K,
acetylene is the most stable among hydrocarbons, although unstable with respect
to carbon and hydrogen. Above 3,000°K, the yield of acetylene at equilibrium
with carbon and hydrogen becomes significant. The oxygen present in coal probably
ends up as carbon monoxide, however, oxygen was not considered. In the presence
of nitrogen, the equilibrium yields of cyanogen and hydrogen cyanide are also
significant. The actual product yields obtained in a practical configuration
such as a plasma arc probably do not reach thermodynamic equilibrium because
of insufficient residence times.

4.3   PYROLYSIS PROCESSES

Pyrolysis is the simplest means of upgrading coal to higher quality fuels. Merely by heating, coal decomposes to gases, tar liquids and char. The gases can be readily burned in an industrial furnace. The tar is the most valuable product because it can be hydrotreated to clean liquid fuels. The char must be utilized in an industrial or utility furnace or gasified to a low Btu or synthesis gas. Because of its low content of volatiles char has poor ignition character- istics and may require special burners or some other means of maintaining flame stability. It can always be burned in a fluidized furnace.

The simplicity of the basic flow sheet of pyrolysis as a coal conversion proc- ess obscures a number of mechanical difficulties that have slowed down its commer- cial development. Chief among these difficulties is the caking and agglomer- ating properties of high volatile bituminous coals upon heating. Those very coals are also the most interesting for their high yield of tar liquids. Rapid heating of coal in a configuration that limits the extent of secondary reactions is another difficulty that has not been entirely overcome. Finally, collecting tar liquids and removing suspended fine solids is also a problem, common to other coal conversion processes. The two processes discussed below have at least par- tially overcome the mechanical problems associated with rapid heating and agglom- eration.

4.3.1   The COED process (refs. 103-105)

The COED process (Char-Oil-Energy-Development) was developed by FMC corporation under contract from the office of coal research (subsequently absorbed into ERDA which in turn was absorbed into the department of energy). The development effort reached the stage of a 36 ton-per-day pilot plant tested in the period 1971-1973. Since then, research and development activities were redirected to a related gas- ification process (COGAS) under private sponsorship.

The central part of the process is shown in the schematic diagram of Fig. 4.21. Coarsely ground coal (<1 mm diameter) is dried and fed to fluidized bed I operating at 600°F with hot combustion gases. The volatiles from I flow to the product re- covery system while the char is carried to fluidized bed II operating at 850°F and subsequently to fluidized bed III operating at 1000°F. Char from stage III is carried to fluidized bed IV (1600°F) where it is partially burned with oxygen. Hot char from stage IV is recycled to stage III to provide part of the required heating. The effluent gases from stage IV provide additional heating as well as fluidizing for stage III. The hot product gases from stage III in turn provide the heating and fluidizing medium for stage II. The gaseous and liquid products from stages I and II are separated. Part of the gases is burned to provide the fluidized medium for stage I, the remainder collected as an end product of the

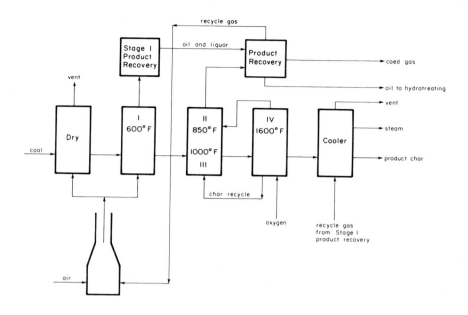

Fig. 4.21. The COED process flowsheet (source: ref. 103).

process. The product liquids are hydrotreated to clean liquid fuels for stationary or portable powerplants.

Carrying out the pyrolysis in four coupled fluidized beds provides the required heating and at the same time prevents caking and agglomeration. As coal
(or char) progresses through stages I-IV its caking temperature increases due to
the successive loss of volatiles. Thus maintaining the temperature of each stage
lower than the caking temperature of the fluidized char prevents its softening
and agglomeration. On the other hand, the prolonged contact of volatiles with
the fluidized char result in extensive secondary reactions. Another adverse
factor in terms of secondary reactions is the relatively large particle size.
Compared to the tars produced in the laboratory reactors described in section 4.1,
the COED liquids are produced at lower yields and have lower boiling points and
less polar character, whence the term "oils".

A ton of high volatile bituminous coal treated by the COED process yields
about 1.4 barrels of oil or about 18% by weight, well below the 30-50% laboratory
yields by the captive sample technique. The other products from one ton of coal
are char, about the same as the ASTM proximate analysis, and 8000-9000 scf of
gases of heating value about 540/scf. The relatively low yield of liquid products

is probably the main reason why the COED process has been dropped from the small list of liquefaction processes scheduled for commercialization.

4.3.2 The Occidental Research Corporation (ORC) process (ref. 106)

The ORC process has been under development since 1969. The original experimental work at the laboratory scale was internally funded. Subsequent work utilized the laboratory unit and a three ton-per-day process development unit which was tested in the period 1976-1978 under contract to the department of energy.

A schematic of the ORC process flowsheet is shown in Fig. 4.22. Coal is pulverized (median size 20-40 µm), dried and pneumatically transported to the pyrolysis reactor where it is mixed with hot recycle char. Solids and gases move cocurrently downward through the reactor and are collected in a cyclone. The solid stream from the cyclone is split in two parts. One part is carried to an entrained flow vessel for partial burning and recycling to the reactor. The remainder is removed as the product char. The volatiles from the cyclone are rapidly quenched and separated into a fuel gas and a liquid product which after hydrotreating provides the main process product.

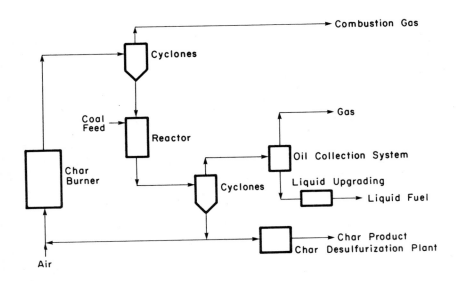

Fig. 4.22. Schematic of the ORC pyrolysis process (source: ref. 106).

Some of the key operating parameters in the reactor are temperature (1000-1400°F), pressure (5 psig) and ratio of recycled char to coal, about 10:1. The results from the bench scale reactor (BSR) and the pilot plant or "process development unit" show some important differences and will be discussed separately.

The bench scale unit consisted of an externally heated entrained flow reactor. Recycle char was not used in this system since the main objective was to determine the dependence of product yields on temperature and residence time. Figure 4.23 shows the yields of the major pyrolysis products as functions of temperature for two residence times, 1.5s and 3s. The most interesting feature of the figure is the maximum in the tar yield at about 1200°F. The maximum yield of about 20% is twice the amount obtained in the standard Fischer assay. As the temperature increases beyond 1200°F the tar yield declines, evidently due to secondary reactions occurring homogeneously or on the reactor walls.

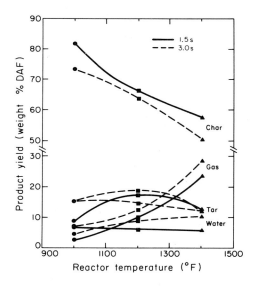

Fig. 4.23. Product yields vs. temperature from the pyrolysis of a subbituminous coal in the BSR (source: ref. 106).

Pyrolyzing the subbituminous coal in the process development unit under similar operating conditions but with hot char recycle gave a surprisingly different product distribution. The tar yield was limited to only about 7-10%, while char and gases were produced in increased quantities. Since the only essential difference between the operation with the BSR and PDU was the presence of the recycle char, the difference in the tar yields was attributed to adsorption and cracking on the char surface. To confirm this explanation additional experiments were performed with increased amounts of recycle char. Changing the char to coal ratio from 10:1 to 40:1 was found to decrease the tar yield from 10% to 4%.

Continuing efforts to improve the tar yield resulted in an unexpected finding. When the inert N_2 carrier was replaced by CO_2 or H_2O, the maximum tar yield from the PDU operation increased to 18-22%, a level identical to that obtained with the BSR. This important and surprising result was attributed to the adsorption of CO_2 and H_2O on the active sites of the char's surface which would otherwise catalyze the cracking of tar molecules. In other words, CO_2 and H_2O compete with tar for the same active sites on the char's surface.

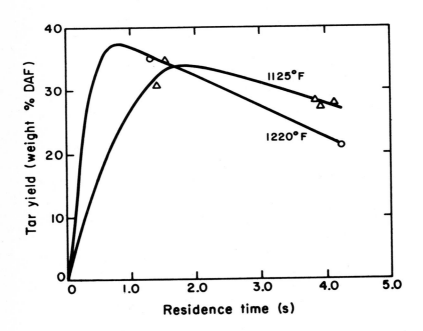

Fig. 4.24. Tar yield vs. residence time from the pyrolysis of a bituminous coal in the PDU (source: ref. 106).

The results discussed up to this point concern the pyrolysis of a subbituminous coal. PDU experiments were also performed with a high volatile bituminous coal (hvc, Kentucky No. 9). Figure 4.24 shows the yield of tar as a function of temperature at two residence times. The maximum yield of about 38% would be quite attractive on a commercial scale. The decline of the yield with increasing residence time is again due to secondary reactions. Surprisingly, the tar yield for the bituminous coal was found to be essentially independent of the carrier gas (N_2, CO_2, H_2O). The strikingly different behavior of the two coals has not been explained and certainly deserves systematic study.

At this point we return to an earlier observation that the caking and agglomerating properties of coal constitute the chief technical difficulty in the commercialization of pyrolysis. The ORC process has approached the difficult problems of coal agglomeration by a special design of the top part of the PDU reactor where the mixing of coal and recycle char takes place (see Fig. 4.25). Pulverized coal is injected by a jet of carrier gas in a surrounding stream of hot recycle char descending downwards from a shallow fluidized region. It is vital that the coal particles are heated rapidly to complete their caking phase before reaching the reactor walls. Small particle size and large char to coal ratio are necessary for this purpose.

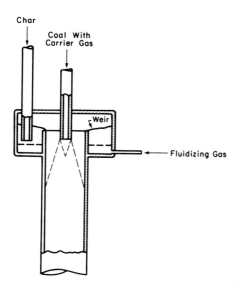

Fig. 4.25. Device for injecting and mixing of coal and char in the PDU (source: 106).

The reactor design shown in Fig. 4.25 was only partially successful. During operation with the bituminous coal the PDU encountered flow instabilities limiting continuous operation to less than one day. A momentary decrease in the char flow rate would cause the deposition of plastic coal particles on the injector tip or the reactor walls. The resulting decrease in the cross section would in turn further reduce the char flow rate and so on, leading eventually to a complete flow stoppage. Although improved designs evolved during the testing, a completely satisfactory operation was not achieved.

Assuming that the mechanical difficulties will eventually be resolved, the ORC process appears a most promising route to coal liquids by virtue of its simple flowsheet and the high yield of tar.

Chapter 5

HEAT AND MASS TRANSFER IN PYROLYSIS

The experimental data reviewed in the previous chapter generally referred to
small particles (<100 μm) and low pressures (atmospheric or lower). Under these
conditions, heat and mass transfer are rapid and have relatively small influence
on weight loss and product yields. In this chapter we will examine the more
general situation in which mass and, to a smaller degree, heat transfer have a
significant influence on product yields. In Chapters 2 and 3 we discussed
chemical structure and reactions with minimal reference to physical properties
such as viscosity and porosity. These properties have a decisive effect on
transport phenomena. Thus in the first section, we briefly discuss the relevant
physical properties of plastic and nonplastic coals. In the second section we
survey experimental data on the effect of pressure and particle size which, of
course, reflect the presence of transport limitations, and in the last section
we develop a simple theoretical treatment of these phenomena. Although heat
and mass transfer are coupled to the kinetics of pyrolysis, the scope of the
theoretical analysis will be limited to problems that can be treated without
reference to detailed kinetics.

5.1 PYROLYSIS AND THE PHYSICAL PROPERTIES OF COAL

5.1.1 The plastic state of coals

The two physical properties that govern the rate of transport processes,
especially mass transfer, are the viscosity during the plastic stage, and the
porous structure of coal. These two properties are not independent because the
plastic properties determine to a major degree the evolution of the porous
structure during pyrolysis. Under certain conditions, when heated above about
$350^{o}C$ coals melt to a highly viscous, non-newtonian liquid, or melt, whence the
term "plastic", or "softening" coals. Whether or not such melting or softening
takes place and the actual viscosity or fluidity (reciprocal of viscosity) of
the coal melt depend on rank, heating rate, particle size, pressure and surrounding gas.

Almost all rheological measurements of coals have been conducted at low
heating rates - a few degrees per minute - using a few grams of coal sample.
Under these conditions, certain coals, chiefly bituminous, become fluid. Fluidity
is most evident in coals with carbon content (dry, ash-free) in the range 81-92%
with a maximum at about 89% (ref. 107). Carbon content does not, of course,
provide complete characterization of rheological properties. The contents of
hydrogen and oxygen are also strongly correlated with fluidity. At fixed carbon

content, fluidity decreases with increasing oxygen content. The latter property
has been already discussed in section 4.2.4.

The conditions of heating play an important role in the development of plastic
properties. At fixed heating rate, increasing pressure and particle size or mass
of sample result in increased fluidity. Particle size and heating rate can, of
course, be varied independently only within a limited range. The plastic properties of dilute pulverized particles depend strongly on the heating rate. Hamilton
(refs. 108, 109) heated dispersed vitrinite particles (100 μm) to $1000^{0}C$ in nitrogen employing heating rates in the range 10^{-1} to 10^{4} $^{0}C/s$ and observed manifesta-
tions of plastic behavior such as the rounding of the particles and the formation
of vesicles and cenospheres. He found a striking relationship between coal rank
and heating rate required for plastic behavior. For high volatile bituminous
coals, changes in char morphology such as rounding, vesiculation, etc. started
becoming significant at about 1 $^{0}C/s$ and increased up to about 10^{2} $^{0}C/s$. Further
increase of the heating rate beyond 10^{2} $^{0}C/s$ did not produce any further morphological changes. Vitrinites of lower or higher rank e.g. subbituminous and
semianthracites required heating rates of 10^{2} $^{0}C/s$ or higher before they displayed
any morphological changes. Once manifested, such changes increased up to 10^{3} to
10^{4} $^{0}C/s$, depending on the particular vitrinite. These results suggest that
coals of different rank can be made to exhibit similar plastic behavior by
suitably adjusting the heating rate.

The plastic behavior of coal and its dependence on rank and heating rate can
be qualitatively accounted by the reactions of pyrolysis. With increasing temperature, the disruption of secondary bonds and the dissociation of covalent bond
induces melting and fluid behavior. The extent of covalent bond breaking required
for this purpose is probably limited, at least for high volatile bituminous coals.
Concurrently with bond breaking, other processes work in the opposite direction
to increase molecular weight and resolidify coal. These are the loss of tar,
which increases the average molecular weight of the remaining material, and the
free radical recombination and various condensation reactions (e.g. condensations
of phenolic groups) which also increase molecular weight. The balance of these
processes determines the occurrence, extent and duration of the fluid or plastic
state. Anthracites are too heavily graphitic in character to exhibit plastic
behavior. Semianthracites and low volatile bituminous coals contain highly condensed aromatic units of relatively large molecular weight. They exhibit some
plastic behavior at sufficiently high heating rates. High volatile bituminous
coals consist of units of lower molecular weight and exhibit maximum fluidity.
With further decreases in rank, the molecular weight of the starting material
would not necessarily decrease, but increased polarity and condensation of phenolic
groups restrict the range and extent of plastic behavior. In particular, plastic

properties are exhibited only at high heating rates. Under such rates, however, fluidity commences at high temperatures and is of short duration due to the acceleration of all reaction rates.

Almost all measurements of rheological properties of coal have been conducted at heating rates of a few degrees per minute. At such rates only bituminous coals exhibit fluid behavior, the fluidity commencing just below 400°C, the maximum fluidity being attained at about 450°C, with resolidification taking place above 500°C. The resolidification relates to the essential completion of tar evolution and the increased molecular weight of the residual material. Coals that exhibit pronounced fluid behavior, are commonly called softening or plastic coals.

The transformation of a coal to a liquid and its subsequent pyrolytic decomposition induce physical changes that have a profound effect on the transfer of pyrolysis products. Following melting, preexisting pores partly collapse due to surface tension forces. The volatile products of decomposition initially dissolved in the melt start nucleating once their concentration exceeds a critical level and the nuclei formed coalesce into larger bubbles which eventually break through the particle surface. Nucleation, growth and bursting of bubbles constitute the chief route of intraparticle mass transfer.

The formation and growth of bubbles causes an expansion or "swelling" of the coal particles. The degree of this swelling depends on particle size, external pressure and heating rate or, generally, temperature-time history. Swelling factors (volumetric) as large as 25 have been observed (ref. 110) under rapid heating conditions. To characterize the swelling properties of coals a standardized test has been developed providing the "free swelling index".

The viscosity of coal in its plastic state has a pervasive influence in many rate processes of interest. It regulates the dynamics of nucleation, bubble growth and bubble coalescence and has an inverse relationship with the diffusion coefficient of pyrolysis products. It also affects the intrinsic kinetics by controlling the rate of bimolecular reactions such as free radical recombination. Viscosity or fluidity (the reciprocal of viscosity) is a transient property and comparisons among different coals are meaningful only under specified conditions of temperature time history, particle size, etc.

Waters (ref. 111) has made extensive rheological measurements suggesting a close relationship between fluidity and instantaneous weight loss. This relation is, of course, due to the fact that fluidity and devolatilization must both be preceded by covalent bond breaking. The rheological properties of coal can be measured by several techniques which have been reviewed in the monograph of Kirov and Stevens (ref. 112). The most common of these techniques employs a rotational viscometer known as the Giesel plastometer, which measures the "fluidity" of coal as a function of time at a heating rate of $3^{0}C$ per minute and other specified experimental conditions.

At heating rates characteristic of flash pyrolysis (several hundred or thousand degrees per second) it is not possible to measure the viscosity, although it is still possible to observe swelling and bubble formation. At high heating rates, the inception of fluidity, the point of minimum viscosity, and the resolidification are displaced towards higher temperatures, in close relation with the rate of weight loss.

An important consequence of coal's plastic properties is the agglomeration of particles to grape-like structures or to a completely coalesced mass or "cake" whence the terms "agglomerating" or "caking" are often used in place of "softening". By contrast, coals which exhibit a limited range of fluid behavior (e.g. sub-bituminous and lignites) are normally considered as "nonplastic", "nonsoftening", "noncaking", or "nonagglomerating". The agglomeration of coal particles is a serious difficulty in the operation of fixed bed or fluidized bed gasifiers and has also been identified as the most serious technical obstacle in the development of commercial pyrolysis processes as a route to coal liquids (section 4.3).

5.1.2 Changes in the porous structure of coal during pyrolysis

The pore structure of coals has been comprehensively treated by Walker and Mahajan (ref. 113) and more recently by Mahajan (ref. 114). These references discuss experimental techniques available for the measurement of surface area, pore volume and pore size distribution. In this subsection we will summarily review the aspects of coal porosity that have an important bearing on transport processes during pyrolysis and the changes of porous structure occurring during pyrolysis.

Coals have a very complex pore structure, both in terms of size distribution, which is very broad, and in terms of the geometry of individual pores or voids. Following ref. 115, we classify pores according to pore diameter into micropores: 0.4 - 1.2 nm, transitional: 1.2 - 30 nm and macropores: 30 - 1000 nm.

Table 5.1 below reproduces measurements of Gan et al. (ref. 115) of pore volumes in the three size ranges for several American coals. The total pore volume V_T was computed from helium and mercury densities, the macropore volume V_1 was estimated from mercury porosimetry, the transitional pore volume V_2 was estimated from the adsorption branch of the nitrogen isotherms and the micropore volume was estimated by difference, $V_3=V_T-V_1-V_2$.

The last two columns in table 5.1 list the surface areas obtained by adsorption of nitrogen and carbon dioxide. S_{N_2} was calculated using the BET equation while S_{CO_2} was calculated using the Dubinin-Polanyi equation. The difference between these two areas has always been of great interest. It is generally attributed to the ability of the carbon dioxide molecule at $298^{\circ}K$ to penetrate pore openings as small as 4 Å, whereas the slightly smaller nitrogen molecule at $77^{\circ}K$ can only penetrate openings larger than about 5 Å.

TABLE 5.1

Pore volumes and surface areas of several American coals (ref. 115).

Designation	Rank	$V_T(cm^3/g)$	$V_1(\%)$	$V_2(\%)$	$V_3(\%)$	$S_{N_2}(m^2/g)$	$S_{CO_2}(m^2g)$
PSOC-80	Anthr.	0.076	75.0	13.1	11.9	7.0	408
PSOC-127	lvb	0.052	73.0	0	27.0	<1.0	253
PSOC-135	mvb	0.042	61.9	0	38.1	<1.0	214
PSOC-4	hvab	0.033	48.5	0	51.5	<1.0	213
PSOC-105A	hvbb	0.144	29.9	45.1	25.0	43.0	114
Rand	hvcb	0.083	47.0	32.5	20.5	17.0	147
PSOC-26	hvcb	0.158	41.8	38.6	19.6	35.0	133
PSOC-197	hvcb	0.105	66.7	12.4	20.9	8.0	163
PSOC-190	hvcb	0.232	30.2	52.6	17.2	83.0	96
PSOC-141	lignite	0.114	19.3	3.5	77.2	2.3	250
PSOC-87	lignite	0.105	40.9	0	59.1	<1.0	268
PSOC-89	lignite	0.073	12.3	0	87.7	<1.0	238

Consider for example sample PSOC-26. The surface area of micropores includes that of pores with openings below 5 Å is $S' = S_{CO_2} - S_{N_2} = 98$ m^2/g. From S' and $V_3 = 0.03$ cm^3/g we can estimate a lower bound for the mean size of the micropores. Assuming spherical shape, the mean diameter of micropores must be at least $6V_3/S' = 1.8$ nm. This estimate suggests that the microporous space largely consists of voids having diameter of a few nm which, however, are accessible via much smaller openings. This particular feature of the microporous system, the "aperture-cavity" structure, has been pointed out by many authors, e.g. refs. 116, 117. Similar observations have been made for cokes (ref. 118) from carbonized coal.

The coals listed in table 5.1 contained substantial porosity in the micro and macro ranges. In particular, micropores constituted more than 60% of total volume in the high rank coals. By contrast, only the high volatile bituminous coals had significant volume in the transitional pore range. Figure 5.1 shows the cumulative pore volume distribution of one such coal (PSOC 190). The distribution covers a wide size range from a few angstroms to about one micron.

The changes in the pore size distribution accompanying pyrolysis depend a great deal on the rank of the coal and, in particular, on its softening or plastic

82

properties. It is thus essential to distinguish between softening and nonsoftening coals. Nsakala et al. (ref. 119) measured the He and Hg densities and the N_2 and CO_2 surface areas of two lignites as a function of isothermal pyrolysis time at 800°C. The lignite particles were injected with a stream of preheated nitrogen in a vertical furnace thus achieving heating rates about 10^4 °C/s. They also measured the He and Hg densities for slow heating (10°C/min) in a fluidized bed maintained at 800°C. At the high heating rates, the helium density increased while the mercury density decreased with pyrolysis time so that the total open pore volume given by

$$V_T = \frac{1}{\rho_{Hg}} - \frac{1}{\rho_{He}}$$

increased with pyrolysis time. The slow heating produced negligible change in the mercury densities but substantial increase of the helium densities. As a result, the helium and mercury densities of the chars produced under slow heating were larger than those possessed by the chars produced by rapid heating. Rapid heating produced sharp increases in the N_2 and CO_2 surface areas as shown for one of the two lignites in Fig. 5.2. In this case, the N_2 surface area increased by a factor of almost one hundred while the CO_2 surface area almost doubled.

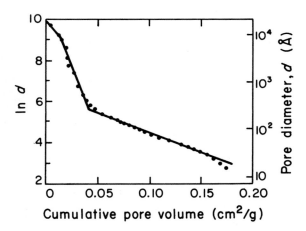

Fig. 5.1. Cumulative pore volume distribution of a hvc bituminous coal "Illinois No. 6" (source: ref. 115).

The authors discussed their experimental results in terms of two competing
processes. Thermal bond breaking produces tar and other volatiles, the removal
of which increases pore volume and widens constrictions, whence the large increase
in the CO_2 surface area. Bond breaking also facilitates the alignment and coalescence of coal's structural units tending to decrease pore volume and surface area.
At the same time, bond formation or cross-linking results in decreased pore
volume and surface area. The balance of these processes depends on coal rank,
heating rate, maximum temperature, and time at the maximum temperature. For
lignites rapidly heated to $800^{\circ}C$, volatile removal predominates over alignment
and cross-linking leading to increased open pore volume and surface area. In
this respect, the increase in the He density in conjunction with the sharp increase in surface area probably signifies the widening of previously impenetrable
apertures. The decrease in the Hg density reflects the removal of material which
at high heating rates is not accompanied by compensating particle shrinkage. At
slow heating, volatile removal is evidently supplemented by cross-linking and
alignment leading to much higher He densities but leaving the Hg densities un-
affected.

Nandi et al. (ref. 120) measured changes in the pore volume and surface area
of three anthracites upon pyrolysis at different final temperatures with heating
rates of $5^{\circ}C/min$. The helium densities in all cases increased with final temperature to about 20% above their initial value. The changes in the mercury

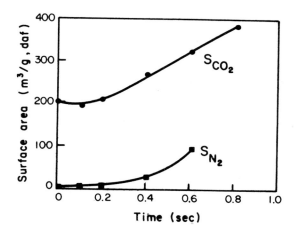

Fig. 5.2. N_2 and CO_2 surface areas of a lignite as a function
of residence time in a vertical furnace. Large times achieved
by multiple passes (source: ref. 119).

densities were smaller and had no definite direction. The change in the total open pore volume was also small and erratic. The N_2 and CO_2 surface areas of two of the three anthracites increased with temperature, passed through a maximum at about $600^{\circ}C$ and then decreased sharply above $800^{\circ}C$. The areas of the third anthracite declined slowly until about $800^{\circ}C$ and rapidly thereafter. The increase in the surface areas at the lower temperatures can again be attributed to the loss of volatiles (e.g. carbon oxides) widening the micropore openings. At the higher temperatures, cross-linking between adjacent units decreased micropore openings causing the sharp decline in surface area.

Toda (refs. 121, 122) studied changes in the pore structure of several Japanese coals following a treatment consisting of heating at $3^{\circ}C/min$ to a final temperature and holding at that temperature for 15 min. The pore structure was probed by mercury porosimetry as well as by measuring the densities in helium, methanol, n-hexane and mercury. Figures 5.3 a, b show the specific volumes in mercury and n-hexane as a function of the final pyrolysis temperature for a nonsoftening and a softening coal respectively. The specific volume in n-hexane is the volume impenetrable by n-hexane, while the volume in mercury is the total particle volume. The difference between the two is the total volume of pores with size above a few angstroms, i.e. it includes macropores, transitional pores, and a portion of the micropores, i.e. those penetrable by the n-hexane molecule. The specific volume in n-hexane of the coals described in Fig. 5.3 and, in fact, of all but one of the coals examined declined with temperature from about $350^{\circ}C$ on indicating consolidation. The decline in this volume is much steeper for the softening coal (Fig. 5.3b), evidently due to melting at about $350^{\circ}C$. The specific volume in mercury for the nonsoftening coal declines monotonically, indicating a shrinkage of the whole particle. In contrast, the specific volume of the softening coal goes through a maximum at about $500^{\circ}C$ indicating mild swelling due to bubble formation, followed by sharp shrinkage signalling the completion of rapid devolatilization and the resolidification of the particles.

An interesting comparison is provided in Figs. 5.4 a, b from the same work of Toda comparing the volume difference $V_{Hg} - V_{n-hex}$ with the total volume of pores above 150 Å as determined by mercury penetration. The close agreement between these two volumes at all pyrolysis temperatures clearly implies that those coals did not possess significant pore volume with openings between a size penetrable by n-hexane (~ 8 Å) and 150 Å. Moreover, no such volume is produced during pyrolysis. The absence of pores with openings between about 8 Å and 150 Å is certainly not a universal property of coals (see table 5.1). Figure 5.4 a, b also shows that the volume $V_{Hg} - V_{n-hex}$ for the softening coal passes through a maximum coincident with bubble formation. This volume, which belongs to pores of size 150 Å or higher, subsequently declines with the disappearance of the

85

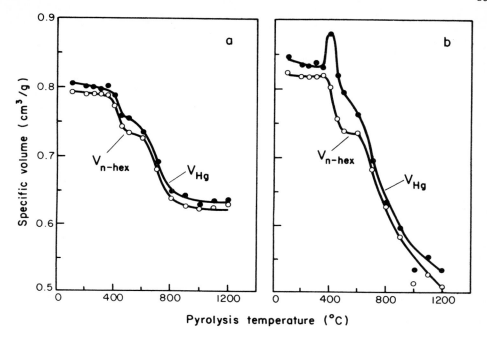

Fig. 5.3. Specific volumes in mercury and n-hexane of a nonsoftening (a) and a softening (b) coal as a function of final pyrolysis temperature at heating rate 3°C/min (source: ref. 121).

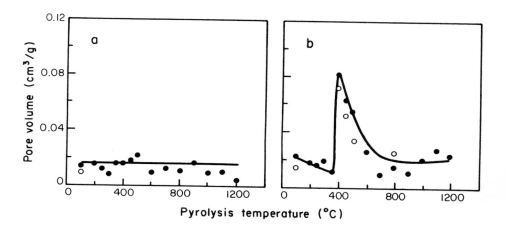

Fig. 5.4. The volume difference $V_{Hg} - V_{n-hex}$ (●) and the volume obtained by mercury penetration (o) for a nonsoftening coal (a) and a softening coal (b) vs. final pyrolysis temperature at heating rate 3°C/min (source: ref. 121).

bubble structure and sustains no further change above 600°C. For the non-softening coal, the volume $V_{Hg} - V_{n-hex}$ shows no significant change with the maximum pyrolysis temperature.

To characterize the evolution of the microporous structure with heat treatment, Toda measured specific volumes in helium and methanol as well as in n-hexane. Figure 5.5 a, b shows these volumes as a function of the final pyrolysis temperature for the non-softening and the softening coals examined in the earlier figures 5.3, 5.4. For both coals the three specific volumes show a rapid decrease after about 400°C suggesting drastic changes in the microporous structure. For the nonsoftening coal, the volumes in methanol and helium reach a minimum at $800 - 900^{\circ}$C above which they increase again. For the softening coal, all volumes

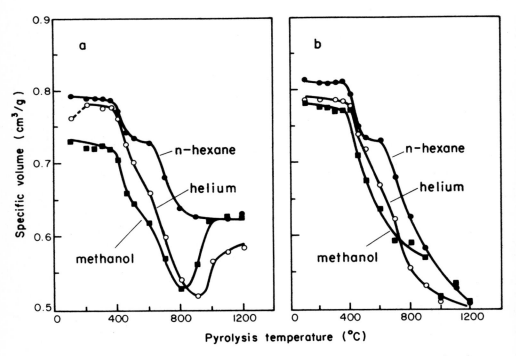

Fig. 5.5. Specific volumes in n-hexane, helium and methanol of a nonsoftening (a) and a softening (b) coal vs. final pyrolysis temperature at heating rates 3°C/min (source: ref. 122).

continue to decline up to the highest pyrolysis temperature utilized (1200°C). The difference between the two types of coal was attributed to the susceptibility of softening coals to alignment and consolidation of their crystallites. In nonsoftening coals, alignment and consolidation are negligible, the primary volume changes being due to widening or narrowing of apertures. Initially, apertures are enlarged due to volatile removal leading to a decrease in the helium and methanol volumes. Beyond 800 or 900°C, apertures start being sealed by cross-linking reactions leading to an increase in the two specific volumes.

Methanol is known to penetrate very fine pores and cause a certain degree of swelling. Because of this penetration, or inbibition, the volume in methanol is lower than that in helium (at least up to 800°C) despite the reverse order in their molecular size. The difference $V_{n-hex} - V_{MeOH}$ was chosen as a measure of the volume of micropores with openings smaller than those penetrated by n-hexane, i.e. smaller than about 8 Å. Figure 5.6 a, b plots the difference $V_{n-hex} - V_{CH_3OH}$ versus the final pyrolysis temperature for the two coals. Both coals display a maximum, the nonsoftening coal at about 700°C, the softening coal at about 600°C. The increasing part of the curves is mainly due to volatile evolution while the

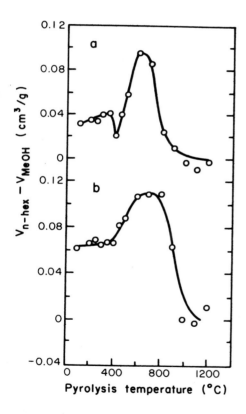

Fig. 5.6. The volume difference $V_{n-hex} - V_{MeOH}$ vs. final pyrolysis temperature for a softening (a) and a nonsoftening (b) coal (source: ref. 122).

88

decreasing part mainly due to sealing of pore apertures by crosslinking reactions. An additional feature of the softening coal is the minimum at about $400^{o}C$, obviously due to loss in pore volume caused by melting.

The results just discussed were obtained with very low heating rates ($3^{o}C$/min). The differences between softening and nonsoftening coals are expected to be more pronounced at higher heating rates which accentuate softening and swelling properties. Unfortunately, very little work has been conducted on the changes in the porous structure of softening coals under conditions of rapid pyrolysis. Figure 5.7 (a) shows the changes in the size distribution of transitional pores of a softening coal heated to $500^{o}C$ at heating rates about $200^{o}C$/s. The main change is the elimination of pores in the range 15 to 60 Å. The total pore volume, however, increased indicating increase in the macropore range due to

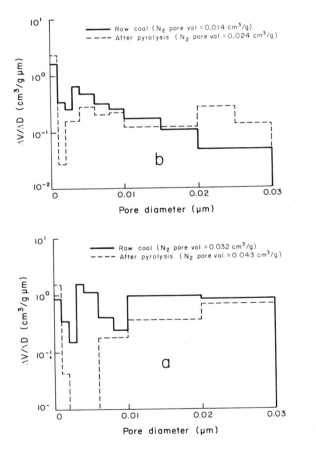

Fig. 5.7. Pore volume distribution of a hvc bituminous coal (a) and a subbituminous coal (b) before and after pyrolysis at $500^{o}C$ for 30 s (source: ref. 62).

bubble formation. At higher temperatures and heating rates, the rapid evolution of volatiles leaves behind large voids occupying almost the entire volume of the swollen particle with the solid matter forming a lacey network of thin shells. A single void occupying almost the whole particle is called a "cenosphere". Photographic studies of these phenomena (refs. 123, 124) have documented the geometry of the large voids but have not given information about changes in the pore volume distribution at the lower end of the size scale.

5.2   THE EFFECTS OF PRESSURE AND PARTICLE SIZE ON PRODUCT YIELDS
5.2.1   The effect of pressure

From the limited data available on the pressure dependence of the weight loss we reproduce here Figs. 5.8 and 5.9 from the work of Anthony et al. (refs. 125, 126). Figure 5.8 shows the weight loss of a hva bituminous coal (Pittsburgh No. 8) as a function of temperature for two pressure levels. Above about 600°C the weight loss at 69 atm is substantially lower than that at atmospheric pressure. Figure 5.9 shows the weight loss at 1000°C as a function of pressure for the same high volatile bituminous coal. Even at atmospheric pressure, the weight loss is substantially below its vacuum value. The difference, however, should be less pronounced at lower temperatures.

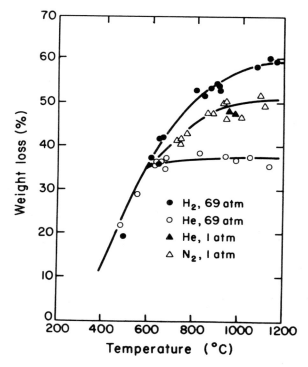

Fig. 5.8.   Weight loss vs. temperature at two pressure levels for a hva bituminous coal (source: ref. 125).

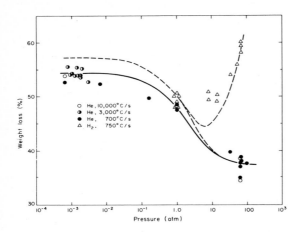

Fig. 5.9. Weight loss vs. pressure at 1000°C for pyrolysis and hydropyrolysis of the Pittsburgh No. 8 coal (source: ref. 125).

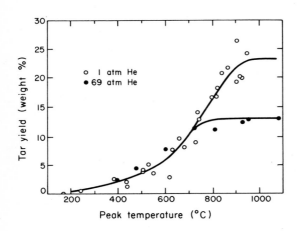

Fig. 5.10. Tar yield vs. temperature at two pressure levels for the Pittsburgh No. 8 coal (source: ref. 63).

An investigation of the pressure dependence of individual product yields for
the Pittsburgh No. 8 bituminous coal was conducted by Suuberg (ref. 63). Figures
5.10, 5.11 summarize some of his results. The tar yield is shown in Figure 5.10
a function of temperature for two pressure levels, 1 and 69 atm. The yields at
the two pressures begin to diverge at about 700°C. At 1000°C the yield at 69 atm
is almost half its atmospheric value. Figure 5.11 plots the yields of various
classes of products versus pressure at 1000°C. The basic trend is very clear.
As the pressure increases the yield of tar decreases while the yields of hydrocarbon gases increase. Since tar is the predominant product on a weight basis
its decrease outweighs the increase in the gases, whence the decrease in the total
volatiles (weight loss).

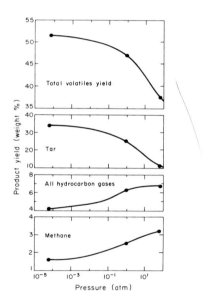

Fig. 5.11. Product yields vs. pressure for the pyrolysis of
the Pittsburgh No. 8 coal in helium at 1000°C (source: ref. 63).

In a study of a hvc bituminous coal Gavalas and Wilks (ref. 62) found a significant pressure dependence of product yields at temperatures as low as 500°C.
The pyrolysis tar of the Kentucky No. 9 was separated by gel permeation chromatography into three molecular weight fractions and it was observed that the vacuum
tar had larger molecular weights than the atmospheric tar (ref. 7). The pressure
dependence of the product yields in the pyrolysis of a lignite was reported by

Suuberg et al. (refs. 63,64). Below 700°C the yields were pressure independent but above that temperature the yield of gases increased with pressure as shown in Fig. 5.12 for the case of methane. In the same figure the yield of other hydrocarbons and tar is observed to decrease with pressure due to the decline of the tar component. The total yield of volatiles, not shown in the figure shows a modest decline with pressure.

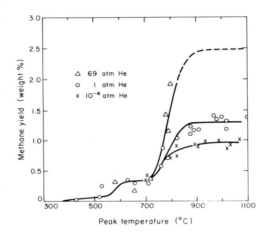

Fig. 12. Methane yield vs. temperature at three pressure levels for the pyrolysis of a lignite (source: ref. 64).

5.2.2   The effect of particle size

Very few experimental data are available concerning the particle size dependence of product yields. Figure 5.13 describing the pyrolysis of a bituminous coal shows that increasing the particle diameter by a factor of ten causes only a modest decrease in the weight loss and the yield of tar and an increase in the yield of gaseous products. A similar effect was found by Anthony et al. (ref. 126) for the Pittsburgh bituminous coal at atmospheric pressure and 1000°C. Gavalas and Wilks (ref. 62) and Solomon (ref. 70) also observed small effects of particle size in the pyrolysis of high volatile bituminous coals in the pressure range vacuum to 2 at and at temperatures 500-100°C.

Significant size effects were observed in the pyrolysis of a subbituminous coal (ref. 62) as shown in Fig. 5.14. The tar yield shows a mixed trend probably due to difficulties in its quantitative recovery. The yield of gases, however, shows a substantial increase with particle size.

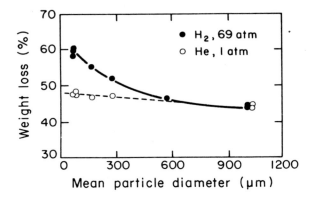

Fig. 5.13. Weight loss vs. particle size for the pyrolysis and
hydropyrolysis of the Pittsburgh No. 8 coal at 1000°C at 1 atm
He. (Source: ref. 125).

In studying particle size as an experimental variable it is necessary to take
some precautions to avoid simultaneous variation of chemical composition due to
maceral enrichment. If a quantity of coal is ground and sieved, the size fractions
will not be uniform in maceral content (and hence chemical composition) because
of the different grindability properties of the macerals. Moreover, the small
size fraction usually becomes enriched in minerals, especially pyrite. The diff-
erentiation in maceral content can be avoided by a simple modification of the
grinding-sieving procedure. The whole coal is ground and a particular sieve
fraction separated representing the largest size desired. A portion of this
fraction is ground again and a second and smaller sieve fraction is separated, etc.

5.3 ANALYSIS OF HEAT TRANSFER

In a broad range of experimental conditions heat transfer is sufficiently rapid
compared to chemical reactions so that it does not influence the weight loss and
yield of individual products. In the first subsection we derive a criterion for
the absence of heat transfer limitations. In the second subsection we briefly
discuss some attempts towards a theoretical analysis of simultaneous heat transfer and kinetics.

5.3.1   Criteria for the absence of heat transfer limitations

To estimate the magnitude of heat transfer effects in coal pyrolysis we shall
examine a very simple model problem that possesses the main features of the real

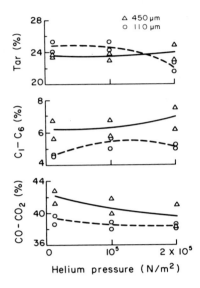

Fig. 5.14. Product yields as percent of weight loss vs. pressure for two particle sizes in the pyrolysis of a subbituminous coal (source: ref. 62).

problem. In the model problem we consider a solid sphere with density, heat capacity and thermal conductivity ρ_s, c_{ps}, λ_s immersed in a hot gas with bulk temperature T_∞. The heat transfer between gas and sphere is described by a heat transfer coefficient h. The interior of the sphere is cooled by a constant and spatially uniform volumetric heat sink of strength bT simulating the heat effect of the pyrolysis reactions. The equations governing this simple situation are

$$\alpha_s \frac{1}{x^2} \frac{\partial}{\partial x}\left[x^2 \frac{\partial T}{\partial x}\right] = \frac{bT}{\rho_s c_{ps}} + \frac{\partial T}{\partial t} \tag{5.1}$$

$$x = 0: \quad \partial T/\partial x = 0 \tag{5.2}$$

$$x = a: \quad \lambda_s \partial T/\partial x = h(T_\infty - T) \tag{5.3}$$

where $\alpha_s = \lambda_s/\rho_s c_{ps}$ and a is the particle radius.

Equations (5.1)-(5.3) can be written in the dimensionless form

$$\frac{1}{\xi^2} \frac{\partial}{\partial \xi}\left[\xi^2 \frac{\partial \theta}{\partial \xi}\right] + \beta(1-\theta) = \frac{\partial \theta}{\partial \tau} \tag{5.4}$$

$$\xi = 0 \quad : \quad \frac{\partial \theta}{\partial \xi} = 0 \tag{5.5}$$

$$\xi = 1 \quad : \quad \frac{\partial\theta}{\partial\xi} = -\lambda\theta \tag{5.6}$$

where $\xi = x/a$, $\tau = \alpha_s t/a^2$, $\beta = a^2 b/\lambda_s$, $\gamma = ha/\lambda_s$, $\theta = (T_\infty - T)/T_\infty$.

The solution of this problem can be written as

$$\theta = \theta_{ss} + \theta_{us} \tag{5.7}$$

where

$$\theta_{ss} = 1 - \frac{\gamma}{\xi} \frac{\exp(\beta^{\frac{1}{2}}\xi) - \exp(-\beta^{\frac{1}{2}}\xi)}{\beta^{\frac{1}{2}}[\exp(\beta^{\frac{1}{2}}) + \exp(-\beta^{\frac{1}{2}})] - (1-\gamma)[\exp(\beta^{\frac{1}{2}}) - \exp(-\beta^{\frac{1}{2}})]} \tag{5.8}$$

$$\theta_{us} = \sum_{n=1}^{\infty} g_n \exp\left[-(\zeta_n^2 + \beta)\tau\right] \frac{\sin\zeta_n\xi}{\xi} \tag{5.9}$$

where g_n depend on the initial particle temperature and $\zeta_1 < \zeta_2 < \ldots$ are the positive roots of the equation

$$\tan\zeta = \frac{\zeta}{1-\gamma} \tag{5.10}$$

As $\tau \to \infty$, θ tends to the steady state profile θ_{ss} for which

$$\theta_{ss}(0) = 1 - \frac{2\gamma\beta^{\frac{1}{2}}}{\beta^{\frac{1}{2}}[\exp(\beta^{\frac{1}{2}}) + \exp(-\beta^{\frac{1}{2}})] - (1-\gamma)[\exp(\beta^{\frac{1}{2}}) - \exp(-\beta^{\frac{1}{2}})]} \tag{5.11}$$

The preceding expressions can be used to investigate the time required to reach the steady state and the shape of the steady state. The length of time is determined by the quantity $\zeta_1^2 + \beta$, the smallest coefficient of τ in the exponentials of Eq. (5.9). This time can be defined by

$$\tau_1 = \frac{1}{\beta + \zeta_1^2} \tag{5.12}$$

showing that the presence of an endothermic reaction speeds up the approach to steady state. The quantity ζ_1^2 is a function of γ(eq. 5.10) which can be estimated as follows

$$\gamma = \frac{ha}{\lambda_s} = Nu \frac{\lambda_g}{\lambda_s}$$

The Nusselt number is usually larger than 2 while $\lambda_g/\lambda_s \sim 0.1-0.5$. Hence $\gamma > 0.2$. An examination of (5.10) reveals that $\gamma > 0.2$ implies $\zeta_1(\gamma) > \zeta_1(0.2) = 0.593$, hence

$$\tau_1 < \frac{1}{\beta + 0.593} \tag{5.13}$$

or in terms of the dimensional time

$$t_1 < \frac{1}{\beta + 0.59} \frac{a^2}{\alpha_s}$$

When the heat effect due to reaction is insignificant, $t_1 < 1.69 \; a^2/\alpha_s$.

We now examine the steady state profile. The quantity $\theta_{ss}(0) = (T_\infty - T(0))/T_\infty$ represents the maximum difference between particle temperature and gas temperature. Eq. (5.11) shows that as $\beta \to 0$, $\theta_{ss}(0) \to 0$ as it should. The net effect of the pyrolysis reactions can be endothermic (b>0) or exothermic (b<0). In a substantial range of experimental conditions $\beta << 1$ and $\theta_{ss}(0)$ can be approximated by

$$\theta_{ss}(0) = \frac{\beta}{\beta+2\gamma} < \frac{\beta}{0.4+\beta} \approx 2.5\beta$$

We take as an example a reaction temperature of $1000^\circ K$, a heat of reaction 30cal/g and $\rho_s = 1 \; g/cm^3$, $\lambda_s = 10^3 \; cal/cms^\circ K$, a=50 µm. To obtain b we need the rate of devolatilization. At $1000^\circ K$ we may assume a devolatilization rate of 3 g/gs so that $b = 30 \times 3/1000 = 0.09 \; cal/cm^3 \; s^\circ K$ and $\beta = 0.005^2 \times 0.09/10^{-3} = 2.25 \times 10^{-3}$. Under these conditions the approximate expression above holds, i. e. $\theta_{ss}(0) \approx 5.6 \times 10^{-3}$ or $T_\infty - T(0) = 5.6^\circ K$.

Absence of heat transfer limitations requires that the time to reach steady state is short, compared to the reaction time, and that the steady state temperature difference is small:

$$\frac{a^2}{\alpha_s} \frac{1}{\beta+0.59} << t_r \tag{5.14}$$

$$2.5 \; \beta << 1 \tag{5.15}$$

These are equivalent to

$$\frac{a^2}{\alpha_s} << t_r \tag{5.16}$$

$$\beta << 0.4 \tag{5.17}$$

The experimental reaction time has the order of magnitude $1/r_0$ where r_0 is the rate of devolatilization in g/gs. Substituting the experimental time for t_r in Eq. (5.14) and using dimensional variables in (5.15) we obtain

$$r_0 \frac{a^2}{\alpha_s} << 0.59 \tag{5.18}$$

$$\frac{a^2 \; r_0 \rho_s |\Delta H|}{T\lambda_s} << 0.4 \tag{5.19}$$

The last equation can be rewritten as

$$\frac{r_0 a^2}{\alpha_s} \frac{|\Delta H|}{Tc_p} << 0.4 \tag{5.20}$$

The ratio $|\Delta H|/Tc_p$ is always much smaller than unity, hence (5.18) implies (5.19). Our final conclusion then is that Eq. (5.18) provides a necessary and sufficient criterion for the absence of heat transfer limitations. It should be kept in mind that r_o is the experimentally measured rate and when Eq. (5.18) is violated, the reaction rate r_o already includes the retarding effects of heat transfer.

As an illustration of the above criteria we may take the rate expression $r_o=$ 870 exp(-13,200/RT) obtained by Suuberg et al.(ref. 71) for the devolatilization of a bituminous coal in the range 1000-2100°K. Curve A in Fig. 5.15 is a plot of the equation

$$\frac{a^2}{\alpha_s}\ 870\ \exp\left(-\ \frac{13200}{RT}\right) = 0.059$$

which defines on the T-a plane the region free of heat transfer limitations.

The above simple analysis can be supplemented to take into account two effects, the radial convective flow in the gas film surrounding the particle, and the swelling of the particle.

The effect of the convective flow in the film can be taken into account (ref. 127) by changing boundary condition (5.3) to

$$x=a:\quad \lambda_s \partial T/\partial x = \frac{Gc_{pg}(T_\infty-T)}{1-\exp(Gc_{pg}/h)} \tag{5.21}$$

When $Gc_{pg}/h \ll 1$ (5.21) reduces to (5.3). Using $G=r_o\rho_s(4\pi a^3/3)/4\pi a^2, h=Nu\lambda_g/a$ we have

$$\frac{Gc_{pg}}{h} = \frac{1}{3Nu}\ \frac{\lambda_s c_{ps}}{\lambda_g c_{pg}}\ \frac{r_o a^2}{\alpha_s} \tag{5.22}$$

Typical values for the first two factors are $\lambda_s c_{ps}/\lambda_g c_{pg} \approx 6$, Nu=2, hence

$$\frac{Gc_{pg}}{h} \approx \frac{r_o a^2}{\alpha_s} \tag{5.23}$$

It is now seen that Eq. (5.18) implies $(Gc_{pg}/h) \ll 1$ and the boundary condition (5.21) reduces to the simpler (5.3).

A more serious complication is encountered in the thermal analysis of strongly swelling coals. Bubble formation in the molten coal increases particle size and decreases density and thermal conductivity. Since the decrease in density outweighs the decrease in conductivity, the thermal diffusivity experiences a modest decrease. The predominant effect, however, is due to the increased particle size. If the swelling factor is known then the heat transfer analysis should be based on the expanded particle rather than the initial particle. The situation, however, is more complex because the swelling factor itself depends on temperature and initial particle size.

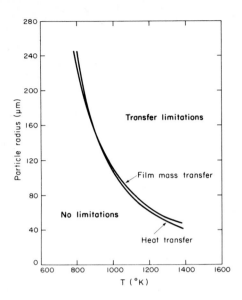

Fig. 5.15. The region of heat transfer (A) and film mass transfer (B) limitations for the pyrolysis rate r_o=870 exp(-13,200/RT)g/gs and α_s=2.2x10^{-3}cm^2/s.

5.3.2. Analysis of combined heat transfer and kinetics

Among the few modeling studies of coal pyrolysis including the effects of heat and mass transfer are those of Mills et al. (ref. 127) and Jones and Mills (ref. 128). The kinetics are described by an empirical set of reactions quite similar to that used in other studies,

$$\text{coal} \xrightarrow{k_1} \text{metaplast}$$
$$\text{metaplast} \xrightarrow{k_2} \text{semicoke + volatiles}$$

The transport aspects of the model are treated in a rather detailed fashion, especially the energy equation which incorporates convective terms due to particle swelling. The swelling is described in an empirical and not altogether satisfactory way, therefore the specific numerical results obtained are not very reliable. Nevertheless, they point out, at least qualitatively, the gross effects of heat transfer limitations. A sample of the numerical results of Mills et al. (ref. 127) is given on Figure 5.17 showing the cumulative evolution of total volatiles as a function of time for different particle sizes. The ultimate weight loss is predicted to be independent of particle size in view of the lack of secondary

99

reactions in the above kinetic model. Hence, the particle size is predicted
to influence only the time required to complete devolatilization.

5.4   ANALYSIS OF MASS TRANSFER

Mass transfer effects are experimentally manifested in the dependence of product
yields on pressure and particle size reviewed in section 5.2. The qualitative interpretation of these effects is straightforward. A molecule generated in the
interior of the coal particle requires a certain length of time before it is removed from the particle. During this time the tar molecule may become reattached
to the condensed phase to eventually decompose into light gases and solid char.
Slow mass transfer thus increases the extent of secondary reactions causing a
higher yield of gases, a lower yield of tar and a lower overall production of
volatiles.

Mass transfer includes two processes, intra-particle and film mass transfer.
Intraparticle mass transfer involves the transfer of a product molecule from its
generation site in the condensed phase to the external surface of the particle.
Film mass transfer involves the transfer of the molecule from the external particle surface to the bulk of the gas.

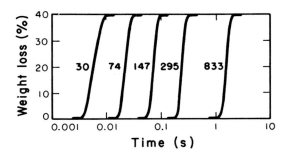

Fig. 5.16. Model calculations of weight loss vs. pyrolysis time for
several particle sizes (μm)(source: ref. 127).

Film transfer proceeds by the same mechanism in softening and nonsoftening
coals although in the former case one might have to consider the change in particle

size due to swelling. The mechanism of intraparticle mass transfer, on the
other hand, differs profoundly between these two types of coals. In softening
coals the transfer of products occurs by nucleation, growth and coalescence of
bubbles during the plastic state of the particles. In nonsoftening coals diffusion
and forced flow through the pore structure constitutes the transfer mechanism.
In the first subsection we examine film transfer seeking criteria for the absence
of transfer limitations. In the second and third subsections we consider intraparticle mass transfer in softening and nonsoftening coals seeking a semiquanti-
tative interpretation of the effects of pressure and particle size.

5.4.1 Film mass transfer

To develop simple criteria for the absence of transfer limitations we consider
a spherical particle immersed in a radially uniform flow field generated by the
flux of products. The products are lumped into two species: gas and tar, the gas
including all low molecular weight products and the tar including all products
with molecular weight above one hundred. The lumping is employed so that we may
use the equations of binary diffusion and thus obtain analytical results. Binary
diffusion at steady state (pseudo-steady state) is described by

$$N_T = y_T(N_T + N_G) - cD_{TG}\frac{dy_T}{dx} \tag{5.24}$$

$$N_G = y_G(N_T + N_G) - cD_{TG}\frac{dy_G}{dx} \tag{5.25}$$

where it is assumed that the only convective term is in the radial direction. In
Eqs. (5.24) and (5.25) N_T, N_G, y_T, y_G are molar fluxes and mole fractions for the
tar and the gas, c is total concentration and D_{TG} is the binary diffusion coefficient. These two equations are equivalent (as shown by addition), therefore only
one need be considered.

In the absence of any source terms in the gas phase the quantities $F_G = x^2 N_T$,
$F_G = x^2 N_G$ are constant and equation (5.24) can be integrated to

$$y_T(a) = v_T(1 - e^{-\sigma}) + e^{-\sigma}y_{T_\infty} \tag{5.26}$$

where $y_T(a), y_{T_\infty}$ are the mole fractions of tar at x=a and x=∞, $v_T = F_T/(F_T + F_G)$ and

$$\sigma = \frac{F_T + F_G}{cD_{TG}a} \tag{5.27}$$

Equation (5.26) can be used to analyze two related but distinct problems. In
the first problem $y_T(a)$ and F_G are considered known and Eq. (5.26) is used to determine F_T. Since v_T and σ are functions of F_T the determination of F_T is implicit.
In a broad range of conditions $\sigma \ll 1$ (see below) and Eq. (26) provides the

approximate solution

$$F_T \approx a D_{TG} c (y_T(a) - y_{T\infty})$$ (5.28)

which after setting $c = p/RT$, $y_T = p_T/p$ can be rewritten as

$$F_T \approx \frac{D_{TG}}{RT} (p_T(a) - p_{T\infty})$$ (5.29)

where R is the gas constant.

Up to this point the analysis is similar to that of Unger and Suuberg (ref. 129). These authors assume that $p_T(a)$ is an equilibrium pressure corresponding to the concentration of dissolved tar, or metaplast, which is *uniform* throughout the particle. The latter assumption probably does not hold in most conditions of interest and will not be adopted in the following.

The effects of pressure and particle size can be explored to some extent without making any statement about the value of $p_T(a)$ which largely depends on intraparticle kinetic and transport processes. Expressing the molar rate of tar production per unit mass of coal we obtain

$$r_T = \frac{4\pi F_T}{(4\pi/3)a^3 \rho_c} = \frac{3}{\rho_c} \frac{D_{TG}}{RTa^2} (p_T(a) - p_{T\infty})$$ (5.30)

where ρ_c is the density of the coal particle. The rate of tar evolution appears inversely proportional to a^2, but this does not take into account the dependence of $p_T(a)$ on particle size. The effect of pressure is twofold. The diffusion coefficient is inversely proportional to pressure while $p_{T\infty}$ is approximately proportional to pressure. Thus as the pressure increases the rate of tar generation decreases resulting in a higher "metaplast" concentration in the coal melt and a smaller cumulative tar yield.

Eq. (5.30) is coupled through the term $p_T(a)$ to the kinetics governing the evolution of the metaplast concentration in the coal melt. Some results can be obtained, however, without kinetic considerations. In experiments carried out by the captive sample technique, $p_T = 0$ because of tar condensation on the cold reactor walls, therefore

$$r_T = \frac{3}{\rho_c} \frac{D_{TG}(p_o)p_o}{RT} \frac{p_T(a)}{pa^2}$$ (5.31)

where $p_o = 1$ atm. This expression implies that the ultimate tar yield depends on the product $a^2 p$ rather than on p and a independently. The scarce data available on particle size dependence do not allow evaluation of this particular result. As we shall see below, however, film mass transfer is in most cases very fast, therefore the pressure and particle size dependence of the tar yield must be mainly due to intraparticle mass transfer processes.

We now return to Eq. (5.26) to derive a criterion for the absence of film mass transfer limitations. Such a criterion may be expressed in the form

$$y_T(a)-y_{T\infty}<<1 \tag{5.32}$$

which can be rewritten with the help of Eq. (5.26) as

$$y_T(a)-y_{T\infty}=(1-e^{-\sigma})(v_T-y_{T\infty})<<1 \tag{5.33}$$

Clearly, this criterion is satisfied when $\sigma<<1$. The parameter σ is a kind of a Peclet number depending on particle size and temperature but independent of pressure since $D_{TG}\alpha 1/p$.

To estimate σ we relate v_T and F_i to the *experimental* rates of tar and gas production:

$$v_T = \frac{r_T/M_T}{r_T/M_T + r_G/M_G} \tag{5.34}$$

$$F_i = \frac{a^3 \rho_c r_i}{3M_i} \qquad i=T,G \tag{5.35}$$

where M_T and M_G are the mean molecular weights of tar and gases, about 300 and 30 respectively. Introducing (5.35) into the definition of σ, Eq. (5.27), we obtain

$$\sigma = \frac{a^2 \rho_c RT}{3pD_{TG}} \left(\frac{r_T}{M_T} + \frac{r_G}{M_G} \right) \tag{5.36}$$

The parameter σ may now be estimated by using the estimates $pD_{TG}\sim(T/T_1)^{1.75}$ atm cm^2/s, $T_1=800^{\circ}$K, $\rho_c=1.2$ g/cm^3, $r_T/M_T+r_G/M_G=r_o/30$. Using the previous expression $r_o=870 \exp(-13,200/RT)$ we obtain

$$\sigma \sim \frac{1.14\ a^2}{T^{0.75}} \exp\left(- \frac{13,200}{RT}\right) \tag{5.37}$$

with a in μm and T in $^{\circ}$K. To obtain concrete results, we arbitrarily set $\sigma=0.1$ as the limit of transfer limitations. The value $\sigma=0.1$ defines curve B in Figure 5.16. This curve lies remarkably close to curve A defining the region of heat transfer limitations. The location of both curves is specific to the rate expression chosen to represent the rate of pyrolysis.

5.4.2. Intraparticle mass transfer in softening coals

We are here primarily concerned with the evolution of tar and gases during the plastic state of coal. In this situation mass transfer consists of two processes in series: diffusion through the molten coal to some internal surface, that of a bubble or a pore; and transport with the bubble or through the pore to the surface of the particle. The role of preexisting pores is not well understood. As discussed earlier in connection with Fig. 5.7, a certain fraction of preexisting pores (<60 Å) collapse during pyrolysis perhaps due to surface tension effects. Pores in the range 60-300 Å were preserved but in this case one could not distinguish between preexisting pores and pores generated by the evolution of bubbles. It appears likely that the major part of mass transfer occurs via bubbles while preexisting pores play a relatively minor role.

A detailed theoretical analysis of mass transfer based on the nucleation, growth and coalescence of bubbles was carried out by Lewellen (ref. 130). The rate of bubble nucleation employed in this analysis depended on several variables including the rate of pyrolysis and contained some adjustable parameters. Bubble growth was described by the Navier-Stokes equations assuming an idealized geometry. Coalescence and bursting were postulated to occur when the surfaces of growing bubbles meet one another or reach the external particle surface. The bubble transport equations were interfaced with the chemistry by assuming that the volatiles experience secondary reactions on the bubble surface. The calculation of extent of these secondary reactions employed bubble surface area and pressure which, in turn, depended in a rather complicated fashion on particle size and external pressure. The analysis thus established the dependency of the volatile yield on particle size and external pressure. The predicted dependence on particle size was in qualitative agreement with experimental data but the predicted dependence on pressure was not satisfactory.

Although adjustments can be made to improve the predicted dependence on external pressure, the coupling between chemical reactions and bubble dynamics remains questionable. As formulated, the analysis does not distinguish between tar and gases and does not take into account considerations of phase equilibrium between bubbles and coal melt which as we shall see below hold the key to the evolution of tar vapors. In spite of these deficiencies and its complexity, Lewellen's analysis contains many useful ideas especially with regard to bubble growth.

To take into account the constraints imposed by thermodynamics we must distinguish between gases and tar vapors. Light hydrocarbon gases, water vapor and carbon oxides have very low solubility in the coal melt at the reaction temperature, therefore they nucleate quite rapidly to bubbles which grow, coalesce and finally burst through the boundary of the molten particle. Tar molecules on the other hand possess relatively low vapor pressure, therefore make a negligible contribution to the nucleation rate. Nevertheless, they do exert a vapor pressure in the bubbles at equilibrium with their concentration in the coal melt adjacent to the bubble interface. It appears very likely that secondary reactions involve primarily the tar dissolved in the coal melt and to a lesser extent the tar vapors in the bubbles or outside the particle.

The partial pressure of tar in the bubbles is at equilibrium with the coal melt at the bubble surface and not necessarily with the bulk of the coal melt. Whether or not substantial gradients of tar concentration exist between the bulk and the interface can be determined, in principle, by comparing the characteristic time for diffusion in the melt with the time for chemical reactions. If D_{TM} is the diffusion coefficient of tar molecules and x_M is the corresponding diffusion path, the characteristic diffusion time may be defined as $t_{DM} = x_m^2/D_{TM}$.

The diffusion coefficient can be crudely estimated from the Stokes-Einstein equation

$$D_{TM} = \frac{k_B T}{6\pi\mu a}$$

where μ is the viscosity of the melt and a the radius of the diffusing molecule. The viscosity of the coal melt is not well defined because of the transient nature of the plastic stage and the presence of bubbles and solid particles. For estimation purposes we may use values in the range $1-5\times10^6$ g/cms (refs. 112,131). Such values correspond to low heating rates (a few °C/min) and temperatures about 420°C. The transient viscosity at higher heating rates and temperatures could be considerably lower. If we use $\mu=3\times10^6$ g/cms and a=8 Å,2.5 Å for tar and gases we obtain the estimates $D_{TM}=2\times10^{-14}$ cm^2/s, $D_{GM}=5\times10^{-14}$ cm^2/s at 500°C. These values are lower by many orders of magnitude than those employed by Attar (ref. 132) and Unger and Suuberg (ref. 129) in their analysis of diffusion in the coal melt.

The diffusion path x_M is the thickness of melt layers between adjacent bubble surfaces and depends on the rates of bubble nucleation, growth and disappearance. Bubble nucleation in coal melts has been analyzed by Attar (ref. 132) but his estimates of nucleation rates are probably too high because of overestimating the diffusion coefficient. Bubble growth, coalescence and disappearance have been analyzed in the aforementioned work of Lewellen (ref. 130). The lack of reliable physical properties and the difficulty of describing bubble coalescence constitute a formidable problem in estimating x_M. However, we can inquire about the values of x_M for which the diffusional resistance in the melt becomes important. For a pyrolysis time of 10s (at 500°C) diffusion starts being important when $t_{MS}=1s$ which corresponds to $x_M=10$ Å, if we use the previously estimated values of D_{TM}. Even if the viscosity were lower by a factor of one hundred, tar transport from the bulk of the melt to the bubble surface would remain one of the rate determining processes. The above observations suggest a qualitative explanation of the role of external pressure in determining the yield of tar. Increasing the external pressure causes the bubble size and the rate of bubble coalescence and bursting through the particle surface to decrease. As a result of the longer bubble residence time and the higher bubble pressure, the partial pressure of the tar increases and tar evolution is slowed down. The result is more extensive secondary reactions and lower cumulative yield of tar.

The tar yield also depends on particle size. Larger particles require larger *internal* bubble pressure to overcome the viscous resistance to bubble growth. As before, larger internal pressure adversely affects the tar yield. However, the particle size effect has not been adequately documented and seems smaller than the pressure effect.

5.4.3.   Intraparticle mass transfer in nonsoftening coals

Mass transfer in these coals proceeds through the porous structure which remains relatively stable during pyrolysis. Much of the methodology of mass transfer in porous solids has been developed in the context of heterogeneous catalysis and is not directly applicable to coal and coal char. The latter materials possess a very broad pore size distribution including micropores 0.0004-0.0012 μm, transitional pores 0.0012-0.03 μm and macropores 0.03-1 μm diameter. This broad size distribution poses two theoretical difficulties: (i) diffusion in the micropores is activated and suitable diffusion coefficients have not been measured at pyrolysis temperatures (ii) coal particles with size of the order of 100 μm, typical of many experimental situations, cannot be strictly treated as continua using an effective diffusion coefficient.

The microporous space of coal and char has molecular sieve characteristics. Diffusion through this space depends on molecular size and is very slow and activated. The pyrolysis products must first slowly diffuse through a domain penetrated by micropores only until they reach the surface of some transitional pore or macropore and then be transported by convection and diffusion to the surrounding gas. The coal or char particle may then be considered as a two phase region, the coal phase, including the micropores, and the void phase consisting of pores larger than 12 Å in diameter. The average diffusion path that a molecule in the coal phase must transverse before reaching the void phase depends on the pore size distribution in a complex fashion which has not yet been investigated. However, for commonly encountered materials some tentative estimates place it in the range 500-2500 Å. The diffusion coefficient of molecules such as CH_4, H_2O, CO etc. in the coal phase is probably larger than 10^{-10} cm^2/s so that under most conditions such molecules require about 1 s before they reach the void phase. Given the low reactivity of these molecules such residence time is too short for any secondary reactions.

Because of their size, molecules in the tar range diffuse much more slowly in the coal phase. Although the diffusion coefficients are not known they could be smaller than 10^{-12} cm^2/s. The corresponding diffusion times are in the range of 100 s, quite long given the propensity of the tar molecules to secondary reactions.

Limitations in the evolution of tar due to this slow diffusion in the coal phase
are independent of particle size (the diffusion path depends on the pore size
distribution and not on the particle size) and cannot be experimentally separated
from the kinetics of pyrolysis. An empirical device for taking into account the
slow diffusion through the coal phase has been utilized in the pyrolysis model of
Gavalas et al. (refs. 133,134).

Once in the pore space (>12 A in diameter) gases and tar vapors are transported
by the customary mechanisms to the particle surface and from there to the bulk
of the gas. The transport of pyrolysis products through the intraparticle void
or pore space depends on particle size and external pressure. This is the dependence that has been studied experimentally and constitutes what is normally
called intraparticle mass transfer. Mass transfer and kinetics are intimately
coupled and should in principle be analyzed together. However, to combine mass
transfer with an elaborate kinetic model requires the solution of a coupled system
of several stiff partial differential equations, a formidable numerical task in-
deed. Russel et al. (ref. 135) studied theoretically mass transfer in combination
with simple empirical kinetics for coal hydropyrolysis. Gavalas and Wilks (ref.
62) obtained some qualitative results concerning the effects of pressure and particle size on pyrolysis yields by formulating a simpler and limited in scope problem. If the instantaneous rates of tar and gas production are known experimentally
an analysis of the mass transfer equations provides the pressure and concentrations
of gases and tar in the pore space. The concentration of tar can then be used as
a qualitative measure of the pressure and particle size effects.

We will now summarize the chief results from reference 62. By assessing the
role of pores of various sizes it is found that pores in the range 300-Å to 1 μm
constitute the main channels for mass transfer. This pore range is approximated
by a bimodal pore system with radii 0.05 and 0.5 μm. The larger pores are few in
number and hence poorly cross-linked, therefore, transport in the particle cannot
be treated using the customary diffusion coefficient and permeability. Instead,
the following simplified assumptions may be employed. Once a molecule enters the
large pores it is quickly carried to the external surface due to the large transport coefficients in these pores. Thus the large pores effectively shorten the
diffusion path through the smaller pores via their mutual intersections. This
situation is approximately described by defining a smaller effective particle
size and considering diffusion through the small pores alone. Binary effective
diffusion coefficients can then be employed.

The multicomponent mass transfer problem can be further simplified by lumping
the various gases into two or three groups according to molecular size. When
pyrolysis is carried out in hydrogen or helium, three components are required:
hydrogen (helium), gases (CO_2,H_2O,CH_4 etc.) and tar. When pyrolysis is carried

out in the atmosphere of pyrolysis or combustion gases, or in nitrogen carrier, two components are sufficient, gases and tar. Although reference 62 treats both the ternary and the binary problem we will here give the results for the binary problem only.

Neglecting accumulation terms, the mass balance equations for gases and tar in a spherical coal particle are given by

$$\frac{1}{r^2}\frac{d}{dr}(r^2 N_i) = \gamma_i \qquad\qquad i = G,T \qquad\qquad (5.38)$$

where N_i are the molar fluxes and γ_i the molar rates of production per unit volume. The source terms γ_i are assumed spatially uniform and known, from the measurements. In a more complete analysis, of course, γ_i would have to be related to the concentrations via the kinetics and would thus vary with the radial position.

With γ_i treated as independent of r, Eq. (5.38) is immediately integrated to

$$N_i = \gamma_i \frac{r}{3} \qquad\qquad i = G,T \qquad\qquad (5.39)$$

The quantities N_i can be described by the approximate flux model

$$N_i = -c_i \frac{\beta}{\mu}\frac{dp}{dr} + N_i^{\sim} \qquad\qquad i = G,T \qquad\qquad (5.40)$$

where β,μ are permeability of the coal and viscosity of the gaseous mixture while c_i are molar concentrations and p is the pressure. In Eq. (5.40) the flux consists of two terms, one due to a pressure gradient and the second, $N_i^{\sim}$, due to diffusion. The diffusive fluxes $N_i^{\sim}$ are given by

$$\frac{dc_G}{dr} = -\frac{N_G^{\sim}}{D_{GK}} + \frac{N_T^{\sim}x_G - N_G^{\sim}x_T}{D_{GT}} \qquad\qquad (5.41)$$

$$\frac{dc_T}{dr} = -\frac{N_T^{\sim}}{D_{TG}} + \frac{N_G^{\sim}x_T - N_T^{\sim}x_G}{D_{GT}} \qquad\qquad (5.42)$$

where D_{ik}, D_{GT} are the effective Knudsen and binary bulk diffusion coefficients given by

$$D_{ik} = \frac{\epsilon}{3} D_{ik}^{*} \qquad\qquad i = G,T$$

$$D_{GT} = \frac{P_{at}}{p}\frac{\epsilon}{3} D_{GT}^{at}$$

where D_{ik}^{*} is the Knudsen diffusion coefficient in a single capillary of the type considered, D_{GT}^{at} is the bulk diffusion coefficient at atmospheric pressure and

p, p_{at} are pressure and atmospheric pressure. The pressure p is a function of location within the particle.

Combining Eqs. (5.39)-(5.42) we obtain

$$\frac{dc_G}{dr} = -\frac{\beta}{\mu}\frac{c_G}{D_{GK}}\frac{dp}{dr} - \frac{r}{3}\left[\gamma_G\left(\frac{x_T}{D_{GT}} + \frac{1}{D_{GK}}\right) - \gamma_T\frac{x_G}{D_{GT}}\right] \tag{5.43}$$

$$\frac{dc_T}{dr} = -\frac{\beta}{\mu}\frac{c_T}{D_{TK}}\frac{dp}{dr} - \frac{r}{3}\left[\gamma_T\left(\frac{x_G}{D_{GT}} + \frac{1}{D_{TK}}\right) - \gamma_G\frac{x_T}{D_{GT}}\right] \tag{5.44}$$

By adding these two equations and using the ideal gas law $c_G + c_T = p/RT$ we obtain an equation for the pressure,

$$\frac{dp}{dr} = -\frac{RT\left(\dfrac{\gamma_G}{D_{GK}} + \dfrac{\gamma_T}{D_{TK}}\right)}{1 + \dfrac{\beta p}{\mu}\left(\dfrac{x_G}{D_{GK}} + \dfrac{x_T}{D_{TK}}\right)}\frac{r}{3} \tag{5.45}$$

By inserting (5.45) into (5.43) and (5.44) the term dp/dr can be eliminated resulting in two coupled equations in c_G, c_T with the initial conditions

$$r = a: \quad c_i = x_{io}(p_o/RT) \qquad\qquad i = G,T \tag{5.46}$$

where p_o, x_{Go}, x_{To} are the known pressure and mole fractions in the bulk of the gas. Equation (5.45) can be written in dimensionless form as

$$\frac{dW}{d\xi} = -\frac{p_{at}}{p_o}\frac{1 + \delta\rho_T}{1 + A(V_G + \rho_T V_T)}B\xi \tag{5.47}$$

$$\xi = r/a, V_i = c_i/(p_o/RT), (i=G,T), W = p/p_o$$

$$\delta = \gamma_T/\gamma_G, \rho_T = D_{GK}^\star/D_{TK}^\star$$

$$A = \frac{3\beta p_o}{\epsilon\mu D_{GK}^\star}$$

$$B = \frac{RTa^2\gamma_G}{\epsilon p_{at}D_{GK}^\star}$$

Although linearly dependent on (5.43) and (5.44), Eq. (5.47) is useful in certain limiting cases where it can be solved independently. For example, in the pyrolysis of subbituminous coals the *molar* production of gases far exceeds that of tar: $\gamma_T << \gamma_G$ and, in addition, $x_{To} << x_{Go}$ and $V_T << V_G$, so that Eq. (5.47) may be simplified to

$$\frac{dW}{d\xi} = - \frac{p_{at}}{p_o} \frac{1 + \delta\rho_T}{1 + AW} B\xi \qquad (5.48)$$

which can be integrated with the initial condition $W = 1$ at $\xi = 1$ to yield the maximum dimensionless pressure difference

$$W(0)-1 = - \frac{1 + A}{A} \left\{ \left[1 + \frac{A_{at}(1 + \delta\rho_T)B}{(1 + A)^2} \right]^{\frac{1}{2}} - 1 \right\} \qquad (5.49)$$

where

$$A_{at} = \frac{3\beta p_{at}}{\epsilon\mu D^*_{GK}}$$

The maximum pressure buildup depends on the dimensionless parameters A, A_{at} and B. For any given coal, the parameter A depends on pressure only and represents the ratio of characteristic times for diffusion and forced flow (due to the pressure gradient). Large values of A imply that forced flow is the predominant mass transfer mechanism while small values of A imply that diffusion is the main mode of mass transfer. The parameter A_{at} is simply the value of A when $p=p_{at}$. The parameter B is a ratio of characteristic times for diffusion and the pyrolysis reaction and encompasses the effect of particle size and temperature. For small values of A (low pressure) transport occurs mainly by diffusion and the intraparticle concentration profiles of tar and gases are determined solely by B.

To examine Eq. (5.49) in more detail, we note that the permeability β for straight cylindrical capillaries of radius R and porosity ϵ is given by $\beta = R^2\epsilon/24$. For $R = 0.05$ μm it can be seen that $A_{at} << 1$. As long as B is not much larger than unity (5.49) can be simplified to

$$W(0)-1 = \frac{1+\delta\rho_T}{2} \frac{A_{at}}{A(1+A)} B \qquad (5.50)$$

or, in dimensional form

$$p(0)-p_o = \frac{1+\delta\rho_T}{2} \frac{A_{at}}{(A/p_o)(1+A)} B \qquad (5.51)$$

The pressure buildup is proportional to the parameter B which is proportional to the square of particle size, and inversely proportional to $1+A$ which includes the effect of pressure. The quantity A/p_o is independent of pressure. At low pressures $A << 1$ and the pressure buildup depends on B alone. At high external pressures, the buildup $p(0)-p_o$ depends on both B and A and diminishes with increasing pressure.

Some results from the numerical solution of Eqs. (5.43), (5.44) are presented in Figure 5.17. The average intraparticle tar concentration is given as a function of A (dimensionless pressure) for three values of B (function of particle size and pyrolysis rate). At low pressures the average concentration is a strong function of particle size but depends very little on pressure. At high pressures the concentration depends on both pressure and particle size. Inasmuch as the extent of secondary reactions depends on the average tar concentration, the above dependence of concentration on pressure and particle size indicates, qualitatively, the effect of these two variables on the yield of tar. To describe this dependence quantitatively requires as we have pointed out the joint analysis of kinetics and mass transfer.

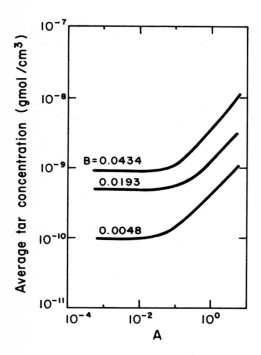

Fig. 5.17. Model calculations of average tar concentration, based on pore volume, as a function of the dimensionless parameters A and B (source: ref. 62).

Chapter 6

KINETIC MODELS OF COAL PYROLYSIS

In this chapter we discuss in detail relatively recent models developed in connection with the experimental programs surveyed in Chapters 4 and 5. Earlier modeling work, now largely superseded, is altogether omitted.

The models discussed in sections 1 and 2 are phenomenological in nature. They
postulate a set of independent or coupled reactions describing product formation
in terms of various components whose exact nature remains unspecified. The postulated reactions contain a number of parameters which are determined by comparison
with experimental data. In some models, the estimated parameter values fall in a
chemically meaningful range and as a result they relate to some extent to actual
chemical processes. Because of their simplicity, phenomenological models have
proved useful in combustion and gasification, where the total yield and heating
content of products are essential while the chemical structure and amount of individual products is not required.

In the last section of the chapter, we will discuss possible approaches to
developing chemical models describing pyrolysis in terms of functional groups and
their elementary reactions. Chemical modeling has not been feasible in the past
because of the complexity of the overall process, the lack of complete grasp on
reaction mechanisms and the lack of sufficiently accurate values for rate parameters. Information accumulating from recent spectroscopic investigations of coal
structure and kinetic experiments with model compounds is gradually improving
this situation to the extent that chemically based models are becoming reasonable
goals in pyrolysis research. Such models would provide the theoretical framework
for further experimentation and a useful tool in process and development work,
especially in the related areas of hydropyrolysis and liquefaction in which the
chemical structure of products and intermediates is of the essence. The models
discussed in this chapter are purely kinetic. Transport processes are either
ignored altogether or are incorporated within the overall kinetics in a rudimentary and empirical fashion.

6.1 INDEPENDENT FIRST ORDER REACTIONS

Although phenomenological models do not employ actual chemical structures and
elementary reactions, they do attempt to provide chemically reasonable kinetic
expressions by postulating hypothetical species participating in simple stoichiometric reactions. In our discussion below, we will attempt to bring
forth the basic premises of each model as well as its utility and limitations.

112

6.1.1 A single first order reaction

This and the next model describe weight loss (or total volatiles) only. The simplest possible description is obviously that of a first order reaction

$$\frac{dW_v}{dt} = - kW_v \tag{6.1}$$

$$W_v(0) = W_v^*$$

where W_v is the weight of volatile precursors in a given sample of coal of initial weight W_o, and W_v^* is the initial value of W_v, which is also the ultimate yield of volatiles, or the ultimate weight loss.

A survey of results using the simple first order kinetics of Eq.(6.1) is included in the review of Anthony and Howard (ref. 136). Three important observations made in this survey are particularly noteworthy. First, the ultimate yield W_v^* usually exceeds the proximate volatile matter (VM) of coal obtained by the well defined ASTM procedure. The ASTM procedure entails extensive secondary reactions inside and on the external surface of the coal particles. By contrast, the pyrolysis experiments of interest here relate to conditions restricting secondary reactions. This difference between the extent of secondary reactions provides an adequate explanation for the discrepancy between W_v^* and VM.

The second observation recalled from ref. 136 is that if the rate constant k is expressed in Arrhenius form,

$$k = A \exp (-E/RT) \tag{6.2}$$

the values of A and E obtained by comparison with the experimental data exhibit large variability. In particular, values of E as low as 4 and as high as 45 kcal/gmole have been calculated while the values of A spanned several orders of magnitude. Some of this variability might be due to the difference in the type of coal but for the most it results from forcing the experimental data into a more or less arbitrary kinetic mold. As discussed in ref. 136, when a set of parallel independent first order reactions is fitted by a single first order reaction, the calculated activation energy can be smaller than each of the individual energies.

The third observation from ref. 136 is that the ultimate weight loss W_v^* often turns out to be a function of temperature, showing in a clear-cut way the deficiency of treating volatile matter as a pseudospecies. Specific results showing the dependence of W_v^* on temperature are included in Figures 4.3-4.5 of the previous chapter.

In spite of the difficulties noted above, Eq. (6.1) is often used for crude estimates and comparisons. For example, the weight loss data from a bituminous coal and a lignite were fitted by the common parameters W_v^* /W_o = 0.7, A = 6.6 x 10^4 s^{-1}, E = 25 kcal/gmole (ref. 57).

6.1.2 Several first order reactions for weight loss

A conceptual improvement in the modeling of weight loss was made by Pitt (ref. 137) who treated coal as a collection of an infinite number of species decomposing by parallel and independent first order reactions. The rate constants were assumed to have common A-factor but different activation energies, varying in a range $[E_{min}, E_{max}]$ according to a probability density function, $f(E)$. Thus $f(E)dE$ is the weight fraction of volatile precursor species with rate constants having activation energies in $[E, E+dE]$. Under isothermal conditions the total weight loss is given by (6.3) where W_v^* is identified as the proximate volatile matter.

$$W_v(t) = W_v^* \int_{E_{min}}^{E_{max}} f(E) \left\{ 1-\exp\ [-t\ A\ \exp(-E/RT)] \right\} dE \tag{6.3}$$

According to this equation, $W_v \rightarrow W_v^*$ as $t \rightarrow \infty$, independently of temperature. However since the pyrolysis time is limited in practice, the calculated ultimate weight loss will be temperature dependent in agreement with the experimentally observed behavior.

Expression (6.3) can be viewed as an integral equation relating the unknown function $f(E)$ to the experimentally measured function $W_v(t)$. Pitt solved this equation by an ingenious approximate technique using weight loss data from the pyrolysis of a high volatile bituminous coal at temperatures 300-650°C and times 10s-100 min. In these calculations A was given the value 10^{15} s^{-1} Because of experimental limitations the weight loss at very short and very long times could not be measured accurately, whence the dotted sections of the curve represent an extrapolation. The weight loss calculated with the curve $f(E)$ of Fig. 6.1 was in good agreement with the experimental weight loss.

It is informative to seek possible chemical interpretation of Pitt's model. As discussed in the previous chapter, tar constitutes 75% or more of weight loss for high volatile bituminous coals, therefore, the sharp peak in Figure 6.1, at about 50 to 55 kcal, would correspond to the tar species. This behavior would be consistent with the hypothesis that the main tar forming reaction involves the dissociation of the ethylene bridge (reaction D5 in chapter 3) whose activation energy would be in the range 48-57 kcal depending on the size and the substituents of the aromatic nucleus. It must be emphasized, however, that it has not as yet been established that the dissociation of the ethylene bridge constitutes the main elementary reaction in tar production. Similar

114

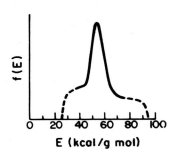

Fig. 6.1. Distribution of activation energies in Pitts' model (source: ref. 137).

interpretation would suggest that the part of the curve at low E corresponds to the formation of H_2O and CO_2 while the part at high E corresponds to the formation of hydrocarbon gases, CO and H_2. For example, reaction D2 which is the main methane forming step has activation energy about 70. The identification of the activation energies involved in the Pitt model with actual bond breaking energies provides the satisfaction that the model reflects some aspects of physical reality.

A further interesting implication of Pitt's model is manifested by attempting to match, in a least squares sense, Eq. (6.3) with the expression pertaining to a single first order reaction

$$W_v = W_v^* \left\{ 1 - \exp\left[-t\, A\, \exp\, (-\overline{E}/RT) \right] \right\} \tag{6.4}$$

The two expressions would not agree very well over a broad range of temperatures, but in a limited range one can compute the value $\overline{E}$ that provides the best fit. The value of E calculated in this fashion turns out to be significantly lower than the mean of $f(E)$. This behavior can explain to some degree the low activation energies determined in reference to the single reaction model.

Going back to Fig. 6.1 it is noted that curve $f(E)$ has a single narrow peak, therefore, it could be well parametrized by two or three parameters, e.g. the location and width of the peak, without seriously compromising the ability to describe experimental weight loss data. Anthony et al. (refs. 126,136) utilized a Gaussian form,

$$f(E) = \frac{1}{(2\pi)^{\frac{1}{2}}\sigma} \exp\left[-\frac{1}{2\sigma^2} (E-E_0)^2 \right] \tag{6.5}$$

containing the two parameters E_0 and σ. They also treated W_v^* as an additional unknown parameter rather than identifying it with the proximate volatile matter. Thus, their model contained four parameters to be estimated from experimental data: W_v^*, A, E_0, σ, only one more than the primitive single first order reaction model. In the original papers of Anthony et al. (refs. 126,136), the limits of integration for Eq. (6.3) were taken $E_{min}=0$, $E_{max}=\infty$. In a later paper (ref. 125), they used E_{min}=0 but treated E_{max} as an additional adjustable parameter. Table 6.1 lists the parameter values determined for two coals. For the lignite they determined two alternative sets which provided equally good agreement with the experimental weight loss data.

TABLE 6.1
Experimental parameter values for the model of Anthony et al.

Coal	Reference	W_v^* (%)	A (s^{-1})	E_o	E_{min} (kcal/gmole)	E_{max}	σ
Lignite	136	40.6	1.07×10^{10}	48.7	0	∞	9.38
"	"	40.6	1.67×10^{13}	56.3	0	∞	10.91
Bituminous	125	37.2	1.67×10^{13}	54.8	0	61.4	17.2

The model of Anthony et al. is sufficiently simple for combustion calculations provided some additional data are available regarding the heating value of the volatiles. An application to fluidized combustion was given in reference 138. In spite its practical success, this model might be criticized for its assumption of independent first order reactions. In later sections of this chapter, this assumption will be discussed in some detail.

6.1.3 Several first order reactions for individual products

In connection with the physical interpretation of the Pitt-Anthony model, it was pointed out that the range of activation energies could be divided into segments which can be tentatively associated with various classes of pyrolysis products. This idea leads quite naturally to describing individual product formation by independent first order reactions. A model of independent first order reactions, one for each product, was employed by the Bergbau-Forschung group (e.g. ref. 74). Suuberg et al. (refs. 64, 71, 72) used a more flexible approach by assuming that certain products can be formed by the breaking of two or more types of bonds, requiring a corresponding number of reactions. The number of reactions required for each product could be judged from the shape of the experimental yield-temperature curves. If each product precursor is recognized as an independent chemical species, its weight W_i changes according to

$$\frac{dW_i}{dt} = - A_i \exp (-E_i/RT) W_i \qquad\qquad (6.6)$$

$$W_i(0) = W_i^*$$

Each product results from one to three reactions, corresponding to different values of i and involving independent sets of parameters (W_i^*, A_i, E_i).

Suuberg et al. (refs. 64, 71, 72) conducted detailed measurements of pyrolysis products for a lignite and a high volatile bituminous coal using the captive sample technique (Chapter 3). In most experiments, the temperature-time history was a pulse consisting of a heating segment with rate about $1,000°C/s$ leading to the desired peak temperature. Immediately after the peak temperature was attained, the current was interrupted and the ensuing cooling segment with rate roughly $- 200°C/s$ completed the pulse. Under these conditions, changing the peak temperature also changed the pulse width, i.e. the effective duration of the experiment. In some experiments, an isothermal temperature segment of 2-10 seconds followed the peak temperature to ensure complete devolatilization. The experiments employing protracted heating were used to determine W_i^*, while the experiments using the simple pulse were used to determine the kinetic parameters A_i, E_i.

Tables 6.2 and 6.3 list the parameter values determined for the lignite and the bituminous coal. Figures 6.2, 6.3 show measured and calculated yields for the lignite, corresponding to the parameters of Table 6.2, while Figures 6.4, 6.5 compare results for the bituminous coal, using the parameters of Table 6.3. The agreement between measurements and model calculations is very good for methane, ethylene and hydrogen (Figure 6.2) which evolve in a stepwise fashion with increasing peak temperature. The agreement is not as good for carbon oxides and water (Figure 6.3) where the experimental points seem to fall on a curve which is gradually increasing rather than stepwise. Within the conceptual framework of the model, the gradual evolution suggests a broad distribution of activation energies that cannot be accurately represented by one, two, or even three discrete values.

In the case of the bituminous coal (Figures 6.4, 6.5) the agreement between model curves and experimental results is good considering the inevitable scatter in the data. The tar yield in this case was calculated using the "evaporative diffusion model" combining kinetics and mass transfer (see Chapter 5). All other products were treated by pure kinetics.

In evaluating the independent reaction model, it should be kept in mind that the data exhibited in Figs. 6.2-6.5 involve total yields for given T-t pulses. No time-resolved data were available. The determination of the rate constants on the basis of cumulative yields becomes increasingly inaccurate with increasing temperature because of a "compensation effect" between the parameters A_i and E_i. In other words, different pairs describe the data equally well, provided a change in A_i is compensated by a commensurate change in E_i. This behavior explains the fact that a common A_i factor for all species other than H_2 and tar described satisfactorily the yields from the bituminous coal. It should also be noted

TABLE 6.2

Experimental parameter values for individual product formation in lignite pyrolysis (ref. 71).

Product	W_i^* (% of coal)	$\log (A_i/s^{-1})$	E_i (kcal/gmole)
CO_2	5.70	11.33	36.2
"	2.70	13.71	64.3
"	1.09	6.74	42.0
CO	1.77	12.26	44.4
"	5.35	12.42	59.5
"	2.26	9.77	58.4
CH_4	0.34	14.21	51.6
"	0.92	14.67	69.4
C_2H_4	0.15	20.25	74.8
"	0.41	12.85	60.4
HC^a	0.95	16.23	70.1
Tar	2.45	11.88	37.4
"	2.93	17.30	75.3
H_2O	16.50	13.90	51.4
H_2	0.50	18.20	88.8

a Hydrocarbons other than CH_4, C_2H_4 and tar

that several of the entries for log A in Table 6.2 are outside the range 12 to 16 pertaining to unimolecular decompositions. Likewise, the activation energy of 37.4 for the first tar forming reaction is much lower than the bond energies estimated in Chapter 3. This is not a serious criticism, however, in view of the great difficulty of multiple-parameter estimation from expressions containing sums of exponentials. The case of tar deserves special attention. The attempt to force the data into a single first order reaction results in a physically unacceptable set of parameters log A = 2.9, E = 13 kcal/mol, in line with the observation made in connection with the single reaction model. To improve on these numbers, the tar yields were fitted (ref. 71) by assuming a set of reactions with continuously distributed activation energies resulting in the parameter estimates log A = 15.4, E_0 = 68.9 kcal/gmole, σ = 11.4 kcal/gmole which

TABLE 6.3

Experimental parameter values for individual product formation in the pyrolysis of a high volatile bituminous coal (ref. 71).

Product	W_i^* (% of coal)	$\log (A_i/s^{-1})$	E_i(kcal/gmole)
CO_2	0.4	13	40
"	0.9	"	65
CO	0.4	"	55
"	2.1	"	65
CH_4	0.7	"	55
"	1.8	"	65
C_2H_4	0.2	"	55
"	0.6	"	65
C_2H_6	0.5	"	55
$C_3H_6+C_3H_8$	0.2	"	40
"	0.8	"	44
"	0.4	"	65
HC^a	0.4	"	55
"	0.8	"	65
HC^b	0.5	"	40
"	1.2	"	55
"	0.4	"	65
Tar	24.0	2.9	13
H_2O	5.4	13	35
H_2	1.0	17	90

a   other hydrocarbon gases
b   light hydrocarbon liquids

are much more in line with the rate parameters estimates of Chapter 3.

The models of Pitt, Anthony et al. and Suuberg et al. are very similar in their underlying assumptions. Both consider product evolution by parallel, independent and first order reactions operating on an initial amount W_i^* of

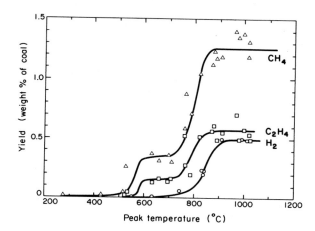

Fig. 6.2   Yields of CH_4, C_2H_4, H_2 from the pyrolysis of a lignite to different peak temperatures at 1 atm He and 1000°C/s heating rate (source: refs. 63, 64).

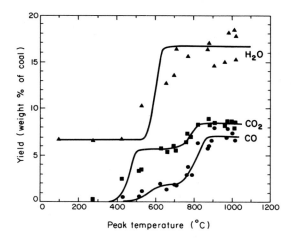

Fig. 6.3.   Yields of H_2O, CO_2, CO from the pyrolysis of a lignite to different peak temperatures at 1 atm He and 1000°C/s heating rate (source: refs. 63, 64).

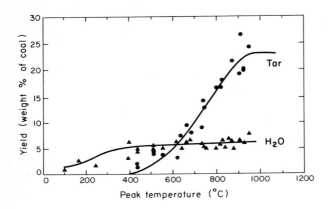

Fig. 6.4. Yields of tar and H_2O from the pyrolysis of a bituminous coal to different peak temperatures at 1 atm He and 1000°C/s heating rate (source: refs. 63, 71).

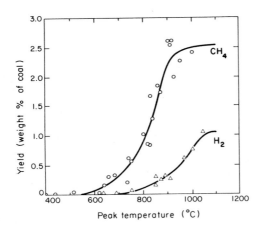

Fig. 6.5. Yields of CH_4, H_2 from the pyrolysis of a bituminous coal to different peak temperatures at 1 atm He and 1000°C/s heating rate (source: refs. 63, 71).

volatile species precursors. When the reactions employed by Suuberg are assembled
to describe the evolution of total volatiles, the collection of activation energies
fall on a bell-shaped pattern, similar to the truncated Gaussian curve of Anthony
et al.

The experimental studies of the Bergbau-Forschung group (ref. 74) employed a
linearly increasing temperature-time history with rates varying from 10^{-2} °C/min
(using an electric oven) to as high as 10^5 °C/min (using an electrical wire-mesh:
the captive sample technique). The rates of production were measured continuously
by mass spectroscopy providing unusually detailed kinetic information. Unfortunately, the production of tar could not be measured in this fashion. As an
example of their results, we consider the production of ethane which they
represented by a single reaction, with log A = 9.33, E = 42.7 kcal/mol, estimated
from the kinetic curve corresponding to a heating rate of 2° C/min. The parameters for heating rates 10^4 °C/min were 9.95, E = 40.9. Although the model
calculations matched the experimental curves quite well (see Fig. 6.6), the difference in the rate parameters (the rate constants differ by a factor of twelve
at 600°C) requires examination. The discrepancy may be due to the inadequacy
of a single reaction to describe the data or to the fact that the temperature-
time history employed is not well suited for kinetic analysis. However, as can
be seen from Figure 6.6 the ultimate yields represented by the areas below the
curves do not differ by more than 15%, therefore, the data are not inconsistent
with the basic premise of independent reactions.

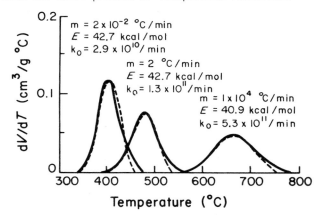

Fig. 6.6. Experimental (----) and calculated
(——) rates of ethane production at different heating rates (source: ref. 74).

We have treated in considerable detail the models of independent first order
reactions because of their attractive features, simplicity and successful

description of experimental product yields. One additional, and in our view important, attribute of these models is the generally reasonable values of the estimated parameters. With some exceptions, the activation energies are in the range of bond dissociation energies associated with functional groups encountered in coal (Chapters 2,3). The idea of product formation occuring by the breaking of specific bonds is straightforward and reasonable.

Undoubtedly, the network of pyrolysis reactions is much more complicated than a set of parallel and independent steps. The models to be discussed in the next section attempt to describe the coupling and competition between reactions leading to different products.

6.2 COMPETING REACTIONS

The consideration of competing reaction schemes has been prompted largely by data showing a negative correlation between the ultimate yields of tar and gases. Under some experimental conditions, an increase in the yield of gases is accompanied by a decrease in the yield of tar and vice-versa. The observed negative correlation could be due to the competition between purely chemical steps or to secondary reactions in conjunction with internal (intraparticle) or external (film) mass transfer limitations.

In chapter 4, we discussed pyrolysis product yields obtained by the captive sample or the entrained flow technique under conditions designed to minimize secondary reactions on the external surface of the particles or the hot surfaces of the equipment. Even under these conditions, however, external secondary reactions may adversely affect the yield of tar at high temperatures and long pyrolysis times. For example, the slight decline in the yield of tar above 800°C in Figures 4.4-4.7 is most likely due to such external secondary reactions. Anthony et al. (refs. 136,126) have pointed out that in most of the early experimental studies, low heating rates were invariably associated with densely packed coal particles, while high heating rates always required dispersed particles. Evidently, the dependence of tar and gas yields on heating rates observed in these studies cannot be attributed to the heating rate per se but rather to the larger or smaller external mass transfer limitations which are incidentally associated with the heating rate.

Intraparticle mass transfer hinders the removal of tar molecules and thereby increases the extent of secondary reactions in the interior of the particles. The extent of such internal secondary reactions increases with pressure and particle size as illustrated by the data presented in chapter 5. Clearly, to describe such mass transfer effects, it is necessary to consider secondary reactions in conjunction with the mass transfer process.

When the pressure is very low and the particles finely ground and well dispersed, the secondary reactions in the interior of the particle are largely supressed. Under these conditions any competition between the yields of tar and gases would be due to purely chemical processes and would be governed by the only remaining operating variable, the temperature-time history. In the experimental work of the MIT group (refs. 63, 64, 71, 72) which devoted particular attention to eliminating secondary reactions, the temperature-time history was varied in three different ways. In the bulk of the experiments, the maximum temperature was varied at constant heating rate . The results of these experiments, which are generally similar to the results presented in Chapter 4, show yields of tar and gas to increase monotonically with temperature, and thus do not indicate the presence of competing reactions. The second type of temperature-time history employed consisted of fixing the maximum temperature while varying the heating rate in the range 250-$10,000^\circ$C/s (refs. 126, 63, 64, 71). For both the lignite and the high volatile bituminous coal that the experiments employed, the yields of tar and gases were within experimental error independent of the heating rate. The third variation of temperature-time history consisted of heating the coal sample to an intermediate temperature, cooling it and then subjecting it to a second and higher temperature pulse (refs. 63, 64, 71). Again, the yields depended on the maximum temperature and not on the details of the heating curve. However, this last type of operation was applied only to the lignite.

The experimental evidence leads to two major conclusions. If mass transfer limitations can be excluded, the product yields depend on the peak temperature and the length of time at that temperature rather on heating rate and other details of the temperature-time history. Future work may reveal some subtler dependencies but the data available to date can be well described by independent reaction models. To describe pressure and particle size effects, it is necessary to consider secondary reactions in conjunction with mass transfer processes. The description of combined mass transfer and kinetics is rather involved as discussed in the previous chapter.

The rest of this section reviews two coupled reaction models. The first model (ref. 129) specifically addressed to softening bituminous coals considers a series-parallel reaction scheme that can be naturally coupled with mass transfer processes. The second model does not include mass transfer limitations but ascribes the competitive evolution of tar and gases to purely chemical processes, namely a set of parallel reactions coupled by common reactants.

6.2.1 Model of Unger and Suuberg (ref. 129)

This model can be represented by the reaction scheme below. First of all, it is assumed that the fraction of coal consisting of the "inactive" macerals (fusinite, micrinite etc.) does not participate in the pyrolysis. The remaining fraction, consisting of the vitrinite, exinite and resinite macerals reacts in the manner indicated below. Water and carbon dioxide evolve by first order reactions (constants k_{i1}, k_{i2}) independent of other subsequent steps. In parallel, coal is converted to an intermediate material called "metaplast," by a first order reaction with constant k_m. The term metaplast was introduced in an early model by Chemin and van Krevelen (ref. 139). The conversion to metaplast proceeds by dissociation of bridges between structural units, therefore, the metaplast molecules, structurally similar to coal, have a distribution of molecular weights. This distribution is assumed to be normal and its mean and standard deviation are treated as adjustable parameters remaining constant during pyrolysis.

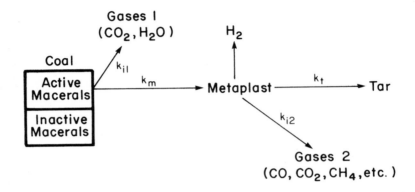

Once formed, metaplast participates in further reactions. The first is polymerization to larger molecules, accompanied by evolution of hydrogen. The effect of this polymerization on the mean and standard deviation of the molecular weight distribution is not taken into account. The metaplast also forms a variety of light products: CO_2, CH_4, C_2H_4, C_2H_6, C_3H_6, C_3H_8, higher hydrocarbon gases and hydrocarbon liquids by parallel, independent reactions with constants k_{i2}. These reactions consume hydrogen to saturate the precursor radicals, hence, their rate is taken to be proportional to the instantaneous content of aliphatic hydrogen in the metaplast.

The final step is the conversion of metaplast to tar. Tar consists of vapors of metaplast fragments, therefore, this step represents the transfer of tar molecules from the condensed phase (metaplast) to the gas phase outside the coal particle. The rate of this mass transfer step is calculated by solving the diffusion problem in the stagnant region surrounding the particle and turns out to be proportional to the vapor pressure of metaplast molecules. The mass transfer problem was discussed in some detail in the previous chapter.

Reference 129 does not include mathematical derivations but gives the parameter values used in the calculations and the comparisons with experimental data. While this model contains a larger number of adjustable parameters, compared to the independent reaction models, it provides a more fundamental description of the pyrolysis chemistry. By virtue of including secondary reactions it allows the description of pressure and particle size effects. The model is a promising one although it requires further work relative to the formation and repolymerization of metaplast and to possible intraparticle mass transfer limitations.

6.2.2 Model of Solomon (refs. 61, 78)

This model describes purely chemical processes and utilizes the following assumptions:

(i) At any instant during pyrolysis, coal is composed of two materials, tar-forming and non-tar-forming, with weights W_T, $W-W_T$, where initially $W=W_o$, $W_T=W_{oT}$. This constitutes the vertical division in Figure 6.7.

(ii) The tar-forming and the non-tar-forming materials are in turn composed of nonvolatile carbon and certain constituents that are precursors to volatile products e.g. H_2O, heavy hydrocarbons, light hydrocarbons, etc. These correspond to the horizontal division in Figure 6.7. The mass fractions of these constituents are the same in the tar-forming and non-tar-forming materials and are denoted by Y_i with initial values Y_{oi}.

(iii) Pyrolysis consists of two types of reactions denoted by DT and DG_i. Reaction DT generates tar molecules by splitting off *vertical* slices containing the instantaneous or current composition of the material. Reactions DG_i produce volatile products by splitting off *horizontal* slices of precursors across both the tar-forming and the non-tar-forming material. Solomon has aptly made the analogy with a bowl of soup being consummed by removing spoonfuls (DT) and by evaporation of ingredients (DG_i).

(iv) The reactions DG_i have the same rates in the tar-forming and non-tar-forming materials, therefore, the mass fractions Y_i remain equal in the two materials. Since Y_i change with time, the composition of evolving tar molecules also varies with time although it continues being identical to the composition in the remaining coal.

(v) The reactions DT and DG_i are all independent and obey first order kinetics. Because the definitions and notation in refs. 61, 78 are somewhat difficult to follow, the derivations given below employ a different notation. The instantaneous weights of the various constituents in the coal undergoing pyrolysis will be denoted by W_i. In particular W_{nvc} will denote the weight of the horizontal slice representing the nonvolatile carbon. This term is somewhat of a misnomer

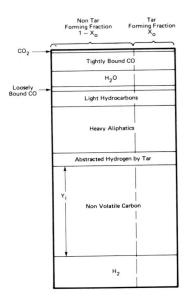

Fig. 6.7. Precursors of various pyrolysis products according to Solomon's model (source: ref. 61).

because although nonvolatile carbon is not affected by any of reactions DG, it is removed as part of the departing tar molecules. The quantity Y_{nvc} $(W-W_T)$ which represents the weight of nonvolatile carbon in the non-tar-forming material is constant,

$$W_{nvc}(W-W_T) = Y_{o,nvc} (W_o - W_{oT}) \qquad (6.7)$$

and proportional to the number of "moles" in that fraction. Likewise, Y_{nvc} $\cdot W_T$ is proportional to the number of "moles" in the tar-forming material. Assumption (v) implies

$$\frac{d}{dt}\left(W_T\, Y_{nvc} \right) = - k_T\, W_T\, Y_{nvc}$$

which under isothermal conditions is integrated to

$$W_T\, Y_{nvc} = Y_{oT}\, Y_{o,nvc}\ \exp (-k_T t) \qquad (6.8)$$

The composition variable

$$y_i = \frac{Y_i}{Y_{nvc}} = \frac{Y_i(W-W_T)}{Y_{nvc}(W-W_T)}$$

is proportional to the "mole fraction" of constituent i in the non-tar-forming material and in the coal as a whole, since the composition of both materials is the same. Assumption (v) then implies that under isothermal conditions

$$\frac{dy_i}{dt} = - k_i y_i$$

therefore,

$$\frac{Y_i}{Y_{nvc}} = \frac{Y_{oi}}{Y_{o,nvc}} \exp (-k_i t) \tag{6.9}$$

Relations (6.7)-(6.9) allow the computation of all quantities of interest. Adding (6.7) and (6.9) yields

$$Y_{nvc} W = Y_{o,nvc} (W_o - W_{oT}) + W_{oT} Y_{o,nvc} \exp (-k_T t) \tag{6.10}$$

therefore, from (6.9)

$$W_i = Y_{oi} (W_o - W_{oT}) \exp (-k_i t) + Y_{oi} W_{oT} \exp [-(k_i + k_T)t] \tag{6.11}$$

The total weight is obtained by summing over i,

$$W = (W_o - W_{oT}) \sum_i Y_{oi} \exp (-k_i t) + W_{oT} \sum_i Y_{oi} \exp [-(k_i + k_T)t] \tag{6.12}$$

where $k_i = 0$ for i=nvc. The composition in the char (remaining coal), $Y_i = W_i/W$ can be calculated from (6.11) and (6.12).

The instantaneous production of constituent i as a gas (and not with tar) is given by $k_i W_i$, hence using (6.11) we find the cumulative production of gas i as

$$\int_0^t k_i W_i dt' = Y_{oi} (W_o - W_{oT}) [1-\exp (-k_i t)]$$

$$+ Y_{oi} W_{oT} \frac{k_i}{k_i + k_T} \left\{1-\exp [(k_i + k_T)t]\right\} \tag{6.13}$$

The ultimate production of gas i (t → ∞) is

$$Y_{oi} (W_o - W_{oT}) + Y_{oi} W_{oT} \frac{k_i}{k_i + k_T} \tag{6.14}$$

Summing over all i≠nvc we find the ultimate weight of gases produced as

$$(1-Y_{o,nvc})(W_o - W_{oT}) + W_{oT} \sum_i \frac{Y_i k_i}{k_i + k_T} \tag{6.15}$$

The ultimate weight of char is obtained from (6.12) as

$$W_\infty = Y_{0,nvc} \; (W_0 - W_{0T}) \tag{6.16}$$

The ultimate weight of tar produced is the difference between $(W_0 - W)$ and the weight of gases produced. Using (6.15) and (6.16) the weight of tar is obtained as

$$W_{0T} - W_{0T} \sum_i \frac{Y_{0i} k_i}{k_i + k_T} \tag{6.17}$$

According to the above expressions the ultimate weight loss is independent of temperature but the ratio of gases to tar varies with temperature due to the relative activation energies of k_i and k_T. Since the activation energies of k_i are generally larger than that of k_T, the ultimate yield of gases generally increases with temperature at the cost of the yield of tar.

The various stoichiometric (X_0, Y_{0i}) and kinetic (k_i, k_T) parameters of the model were estimated by comparisons with experimental data and the estimated values are given in ref. 61. Although the procedure used in the estimation was not explained in detail, it probably involved some trial and error in as much as expressions (6.14) giving the yield of gaseous products include stoichiometric as well as kinetic parameters. The agreement between model predictions and experimental data may be judged from Figures 4.4 - 4.11 where the solid lines represent model calculations. The agreement is good in most cases, however, because of the scatter in the data, the comparisons do not provide a critical test concerning competitive product evolution and other assumptions in this model.

In comparing the models of Solomon (ref. 61) and Suuberg et al. (refs. 64, 71, 72) we note that both employ essentially the same number of adjustable parameters. Although Solomon's model appears more flexible allowing for the competitive evolution of gases and tar, its ability to represent experimental data does not appear to be superior to that of the independent reaction model. The distinction between the two classes of reactions DT and DG_i is useful from the theoretical standpoint, however, the a priori division between tar-forming and non-tar-forming materials of identical composition is clearly empirical. The estimated model parameters include very low activation energies for reaction DT and several of the DG_i reactions. This may again arise from representing several reactions of different activation energies by a single reaction.

6.3 DETAILED CHEMICAL MODELS

In principle, a detailed chemical model should describe pyrolysis in terms of coal's functional groups and their elementary reactions. Chapters 2 and 3 gave a survey of structural and kinetic information that might be useful for this purpose. However, detailed chemical modeling remains hampered by many serious difficulties. Available structural information continues being tentative and qualitative. For example, the relative concentrations of methylene, ethylene and ether bridges, constituting one of the most important structural features, have not yet been quantified. A second difficulty is the large number of individual elementary reactions making the mathematical analysis very unwieldy. A third yet difficulty is the lack of reliable physical properties needed for characterizing mass transfer within the pyrolyzing particles. In view of these difficulties the efforts to develop a reliable model based on functional groups and chemical reactions have made very limited progress. In the first section of this chapter we outline a recent kinetic model by Gavalas, Cheong and Jain (refs. 65, 133, 134) and in the second section we discuss possible directions for future kinetic modeling

6.3.1 A detailed kinetic model (refs. 65, 133, 134)

This chemical model rests on the following general assumptions:

(i) Coals belonging to a broad range of rank, e.g. subbituminous or bituminous, can be characterized by a common set of functional groups, with different coals differing by the concentrations of these functional groups.

(ii) At high temperatures the functional groups react by well known free radical mechanisms.

Since most functional groups are substituents to aromatic nuclei their reactivity depends on the type of nucleus and the existence of other neighboring substituents. For example, the two bond dissociation reactions

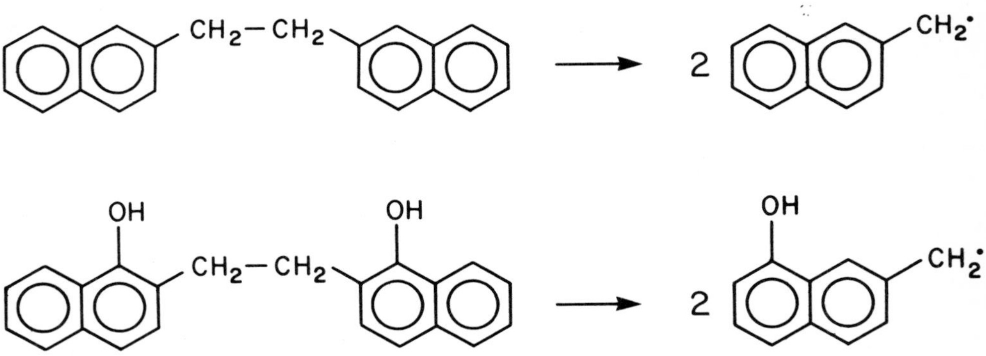

$$\text{naphthyl–}CH_2\text{–}CH_2\text{–naphthyl} \longrightarrow 2\ \text{naphthyl–}CH_2^{\bullet}$$

$$\text{(OH)naphthyl–}CH_2\text{–}CH_2\text{–naphthyl(OH)} \longrightarrow 2\ \text{(OH)naphthyl–}CH_2^{\bullet}$$

involve the same group, the ethylene bridge. The second reaction, however, is much faster than the first on account of the presence in the ortho position of the activating hydroxyl group. A complete description of the kinetics must distinguish between those ethylene bridges that are associated with hydroxyl groups in the ortho position and those that are not. A description at this level of detail would require consideration of a very large number of chemical structures and would be completely out of line with experimental information. To simplify the situation, the following additional assumption is made:

(iii) Any given functional group is exposed to an average environment of other groups, therefore, its reactivity depends only on the *concentration* of other groups.

Assumption (iii) is subject to certain exceptions that can be illustrated by the following example:

$$\phi\text{-}CH_2\text{-}CH_2\text{-}\phi' \rightarrow \phi\text{-}CH_2\text{\textbullet} + \phi'\text{-}CH_2\text{\textbullet} \tag{a}$$

$$\phi\text{-}\overset{\text{\textbullet}}{C}H\text{-}CH_2\text{-}\phi' \rightarrow \text{no dissociation} \tag{b}$$

The dissociation of the ethylene group is completely suppressed in the presence of another group, the α radical. In this case it is necessary to calculate the fraction of ethylene bridges that *do not carry* an α radical. Similar restrictions apply in the presence of double bonds. Such restrictions, or constraints, are taken into account by calculating the configurations of functional groups that are susceptible to reactions. The concentration of these *reactive configurations* is calculated using another yet simplifying assumption:

(iv) At any instant, the functional groups are located completely randomly, subject to some obvious constraints (e.g. no two radicals or radical and double bond can coexist on a single carbon atom). To calculate the concentration of reactive configurations in terms of the concentrations of functional groups requires the solution of a combinatorial placement problem which is solved by standard techniques akin to those used in statistical mechanics.

Once the concentration of reactive configurations has been calculated, the rate of generation of all volatile products other than tar can be calculated. To calculate the rate of tar production, certain additional constraints must be considered. Tar molecules contain, by definition, one or more structural units (see chapter 2), therefore, one or more aromatic nuclei. The vapor pressure and diffusion coefficients of these molecules in the condensed phase decrease rapidly with increasing molecular weight. Thus tar precursor molecules (metaplast) generated in the condensed phase have very small probability of evolving to the gas phase unless they contain only one structural unit.

Having placed a molecular size constraint, we note that the rate of generation of tar molecules is the product of three factors. The first is the concentration of precursor units *connected by a single bridge* to the coal phase. The concentration of precursor units is calculated by a combinatorial placement technique similar to the one employed for reactive configurations. The second is the dissociation rate of the bridges. The third factor is the probability that a unit that has become free by bridge dissociation will be transferred to the gas phase, rather than participate in further reactions in the coal phase.

The last factor is quite complicated as it involves mass transfer processes in the condensed phase and the pore space. Considering first the condensed phase, a free unit produced by a dissociation reaction such as

$\phi\text{-CH}_2\text{-CH}_2\text{-}\phi' \rightarrow \phi\text{-CH}_2\text{·} + \phi'\text{-CH}_2\text{·}$

is subject to immediate recombination by the reverse reaction (cage effect) unless it can quickly transfer to the pore space. Likewise, a free unit formed by

$\text{H·} + \phi\text{-CH}_2\text{-}\phi' \rightarrow \phi\text{-CH}_2\text{·} + \phi'\text{H}$

must diffuse to the surface of a pore or a bubble before escaping to the gas phase. During this diffusion process, the unit is subject to recombination with free radical sites on the coal matrix. Because these phenomena are very difficult to quantify, the model introduces an empirical adjustable parameter X describing the probability that a free unit formed by dissociation reactions (metaplast molecule) will be transferred to the gas phase rather than recombine in the coal phase. In some computer calculations X was given values in the range 0.1-0.2. Once in the pore phase, a free unit was assumed to transfer outside the particle with no further diffusional limitations. This very crude way of handling intraparticle mass transfer fails to account for the effects of pressure and particle size discussed in chapter 5.

Having defined the main physicochemical assumptions we can summarize the model by the block diagram of Fig. 6.8. Blocks 1,2 define the functional groups whose concentrations characterize the state of the coal at any instant during pyrolysis. Some sample calculations discussed in ref. 102 employed fifteen functional groups including two radicals, alpha and beta. Block 3 includes the calculations of the concentrations of reactive configurations referred to previously. The information about reaction mechanisms and rate parameters is introduced through block 4 which in combination with block 3 derives expressions for the rates of individual reactions (block 5). Block 6 includes the calculations of the rate of tar production. The rate of change of any given functional group consists of a direct term, due to one or more elementary reactions, and an indirect term due to the loss of tar molecules carrying the functional group under question. These two terms are combined in the differential equations describing the evolution of the state variables (block 7).

Figure 6.9 is a sample of model calculations compared with a few experimental points for a high volatile bituminous coal. The kinetic parameters used in the calculations were specified as follows. The initial concentrations of functional groups were estimated from elemental analysis and nmr data of pyrolysis tars. Most kinetic parameters were taken from the literature or estimated by thermochemical kinetics while a few parameters were adjusted to get reasonable agreement with experimental data. The adjusted values were in all cases within physically reasonable limits.

Using this set of parameters as a base case, sensitivity calculations were carried out varying one parameter at a time and recomputing the product yields.

132

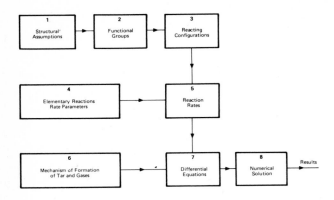

Fig. 6.8.   Organization of the model of Gavalas et al.
(refs. 133, 134).

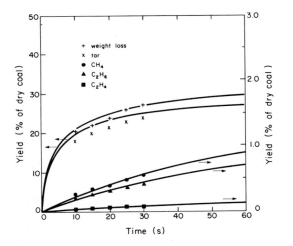

Fig. 6.9.   Calculated (———) and experimental product yields
from the pyrolysis of a hvc bituminous coal, "Kentucky No. 9,"
at 510°C and 1 atm He (source: ref. 134).

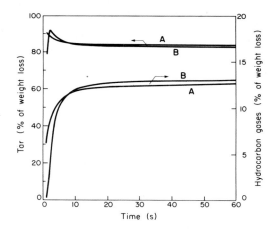

Fig. 6.10. Calculated effect of temperature-time history on relative product yields from the pyrolysis of an hvc bitumi- nous coal, "Kentucky No. 9," at 1 atm He. A: isothermal at $600°C$; B: $400°C$ to $600°C$ linear in 2s followed by isothermal at $600°$ (source: ref. 65).

The parameter with the greatest effect on the results was the rate constant for bridge dissociation (reaction D5, Ch3). Other important parameters were the average number of bridges per structural unit (extent of cross linking) and the rate constants for radical dissociation reactions (DB1-DB4, chapter 3).

To test the effect of temperature-time history, calculations were performed for two different conditions. One was isothermal pyrolysis at $600°C$. The other consisted of a period of two seconds of linearly rising temperature, from 400 to $600°C$, followed by isothermal operation at $600°C$. The isothermal opera- tion resulted in a slightly higher weight loss (about 2.5%), slightly higher yield of tar and slightly lower yield of gases. Figure 6.10 compares relative product yields under the two temperature-time histories. The difference in the relative yields is slight, especially for tar. Before drawing more general con- clusions, calculations should be made for several other types of temperature- time histories.

In view of the use of several adjustable parameters, the comparison between calculated results and the few experimental points (Fig. 6.9) have very little significance, other than indicating the plausibility of the reaction scheme and pointing out the main pathways for the production of various products. As formulated in refs. 133, 134 the model suffers from two serious defects. One is the complexity of the rate expressions resulting from the combinatorial calculation of reactive configurations. The other is the failure to simplify the reaction network by eliminating a number of relatively unimportant functional groups and reactions. The possibilities for such simplifications have emerged from recent structural and model compound studies but have yet to be exploited in kinetic modeling. The next subsection will thus be limited to a qualitative discussion of these possibilities.

6.3.2 Further ideas on kinetic modeling

Recent structural and kinetic studies on whole coal, coal-derived liquids and coal-like model compounds have provided a sharper focus on the functional groups and chemical reactions most significant in pyrolysis. This newer information suggests a revision and simplification of the list of functional groups and reactions employed in refs. 133, 134. By restricting the number of functional groups, it is no longer essential to utilize random distributions, hence, the rate expressions can be considerably simplified. At the same time the hydroaromatic structures attain a more crucial status in the overall reaction network. The following is a revised list of functional groups and chemical reactions that might be suitable for future modeling efforts.

functional groups

To attain a manageable model it is essential to keep the number of functional groups at a minimum. It is thus almost necessary to consider an average aromatic nucleus, common to all structural units. The average nucleus would contain a certain number of aromatic carbons and hydrogens and heteroaromatic oxygen sulfur and nitrogen. These numbers are not integers because they represent not any particular ring system, but an average ring system.

The substituents on the aromatic nucleus may be limited to methyl and phenolic hydroxyl. The exclusion of aliphatic chains longer than methyl is suggested by the work of Deno and coworkers (refs. 12-14) but may be inappropriate for certain coals.

Hydroaromatic structures are very important and must be carefully considered. The four groups below have been tentatively suggested in recent structural studies.

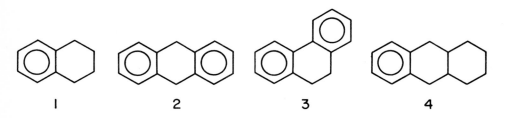

| 1 | 2 | 3 | 4 |

These groups have not been directly identified but have been inferred from the interpretation of various analytical data. The selective oxidation studies of Deno and coworkers suggest the absence of unsubstituted tetralin structures, although they allow for the possibility of methyl-substituted tetralin structures (ref. 12). Likewise, structure 2 was suggested as predominant in an Illinois No. 6 coal, while structure 3 was only indicated as possible. Structure 4 has been employed in order to explain the relatively low ratio of aliphatic hydrogen to aliphatic carbon determined from 'H and ^{13}C nmr spectra of coal extracts (ref. 140). Other complex multiring structures can also be employed for this purpose Some recent unpublished functional group analysis of coal derived liquids by this author also strongly suggests the presence of complex hydroaromatic structures like 4.

The final type of groups are the bridges. Most recent publications suggest methylene ($-CH_2-$) and diaryl ether ($-O-$) as the most abundant bridges between aromatic nuclei. In this respect it must be noted that structure 2 above can be considered either as a hydroaromatic structure or as two methylene bridges. However, in the process of pyrolysis it is unlikely that both bridges could be dissociated before other reactions take place. Along the same line of thought, the biphenyl bond is not meaningfully considered as a bridge because it does not dissociate by thermal means. Finally, ethylene bridges $-CH_2-CH_2-$ have not been specifically identified or excluded on the basis of analytical data.

chemical reactions

The volume of evidence favors free radical rather than pericyclic (concerted) mechanisms, although the latter may play some limited role. One reaction that probably occurs by a non-radical mechanism is the condensation of phenolic groups.

Limiting attention to free radical mechanisms we first consider the set of propagation reactions which determine the relative yield of various products. Prominent among these are the reactions of hydrogen atoms and methyl radicals. The hydrogen atoms participate in two types of reactions, hydrogen abstraction and addition illustrated by the examples

$$H \cdot + PhCH_3 \rightarrow H_2 + PhCH_2 \cdot \qquad (r1)$$

$$H \cdot + PhCH_3 \rightarrow PhH + CH_3 \cdot \qquad (r2)$$

$$H \cdot + PhCH_2Ph' \rightarrow PhCH_2 \cdot + Ph'H \qquad (r3)$$

Reactions (r2) and (r3) have already been discussed in section 3.6. They are important steps for bridge dissociation (r3) and methane formation (r2). Methyl radicals react quite similarly, e.g.

$$CH_3 \cdot + PhCH_3 \rightarrow CH_4 + PhCH_2 \cdot \qquad (r4)$$

$$CH_3 \cdot + PhCH_2Ph' \rightarrow PhCH_2 \cdot + Ph'CH_3 \qquad (r5)$$

Alpha radicals participate in hydrogen exchange reactions such as

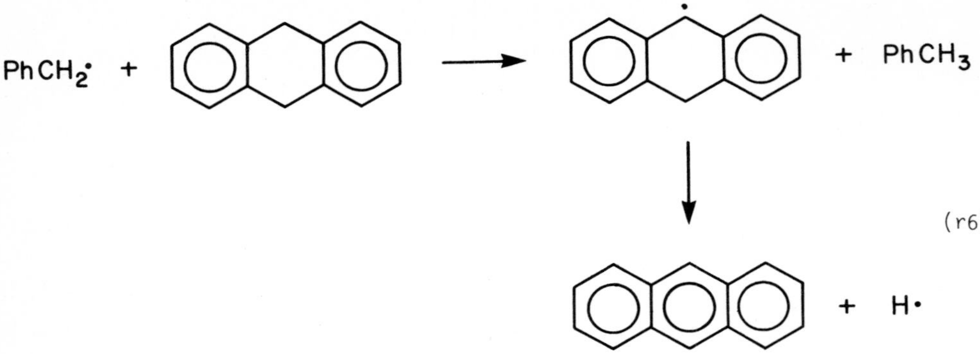

$$PhCH_2^{\bullet} \;+\; \text{[anthracene-dihydro structure]} \longrightarrow \text{[anthracenyl radical structure]} \;+\; PhCH_3$$

(r6)

$$\downarrow$$

$$\text{[anthracene structure]} \;+\; H\bullet$$

Reaction (r6) serves to saturate alpha radicals and at the same time regenerates hydrogen atoms. While the hydroaromatic groups 2,3 react fairly simply as in (r6) above, the groups 1,4 react in a more complicated way. The propagation reactions of group 1 (tetralin-like structure) are summarized below, where the phenyl ring could actually be a naphthyl or other ring system. Reaction (r7) leads to methylindan while (r8) leads to the relatively unstable dihydro-naphthalene which eventually ends up as naphthalene (r10). The addition reaction (r_{11}) can be followed by various hydrogen abstractions and dissociations

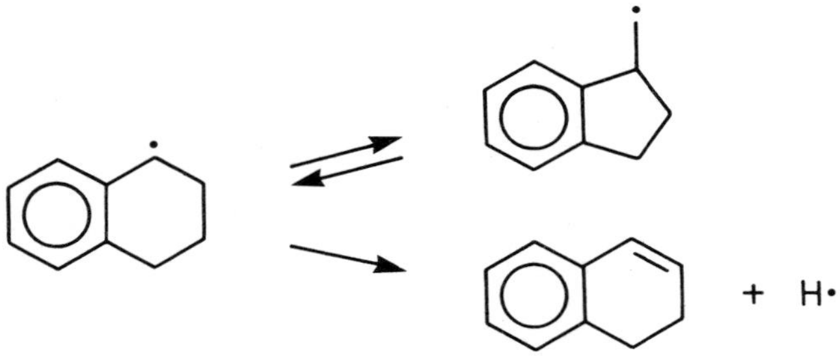

(r7)

$$\text{[tetralinyl radical]} \;\rightleftharpoons\; \text{[methylindan radical structure]}$$

$$\longrightarrow\; \text{[dihydronaphthalene structure]} \;+\; H\bullet$$

(r8)

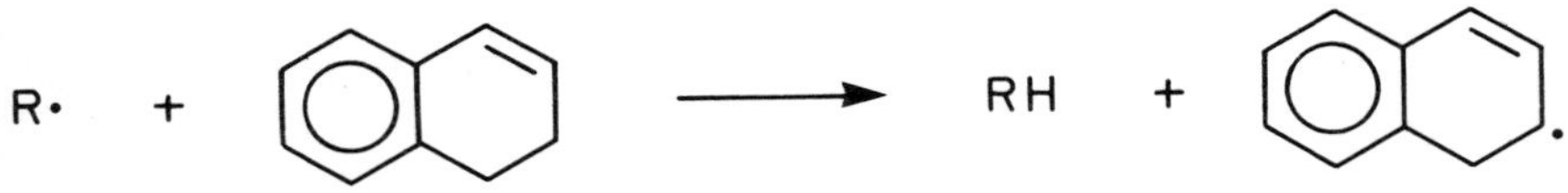

$$R\bullet \;+\; \text{[dihydronaphthalene structure]} \longrightarrow RH \;+\; \text{[dihydronaphthalenyl radical structure]}$$

(r9)

$$\text{[dihydronaphthalenyl radical structure]} \longrightarrow H\bullet \;+\; \text{[naphthalene structure]}$$

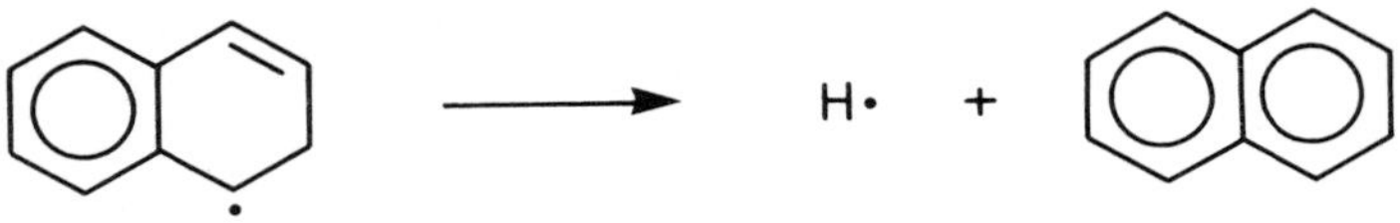

(r10)

$$H\cdot \quad + \quad \text{(structure)} \quad \longrightarrow \quad \text{(structure)} \qquad (r11)$$
$$\text{(or } CH_3\cdot)$$

$$\text{(structure)} \quad \longrightarrow \quad \text{(structure with R)} \quad + \quad (C_2H_4, C_2H_6, C_4H_{10}) \quad (r12)$$

resulting in C_2-C_4 hydrocarbons. The point to be emphasized here is that light
hydrocarbon gases other than methane need not arise from aliphatic chains but may
largely arise from the decomposition of hydromatic structures as illustrated by
reactions (r11), (r12).

Among various possible initiation reactions, only three are relatively energetically favorable,

$$Ph-CH_2-CH_2-Ph' \rightleftarrows Ph-CH_2\cdot + Ph'-CH_2\cdot \qquad (r13)$$

$$\text{(structure)} \rightleftarrows \text{(structure)} \longrightarrow \text{(structure)} + C_2H_4 \quad (r14)$$

$$\text{(structure)} \rightleftarrows \text{(structure)} \longrightarrow \text{(structure)} \qquad (r15)$$

where the phenyl ring could be any aromatic ring system. Which of reactions (r13-
r15) is more important is not known at this time because of uncertainties about
the concentration of ethylene bridges and the magnitude of the cage effect on the
rate of (r13). The kinetics of reactions (r14), (r15) are also poorly understood
as they involve ring opening with formation of biradicals. As a result of these
uncertaintaies the rate of initiation reactions cannot at this point be predicted
from first principles, even if complete structural information was available.

Termination reactions are due almost entirely to recombination of the relatively abundant alpha radicals. The rate of this recombination is controlled by
diffusion and might be negligible during the period of rapid product formation.

138

Chapter 7

HYDROPYROLYSIS

Heating coal in hydrogen rather than in an inert gas results in a significantly
different product distribution and merits separate consideration. In particular,
the increased production of single ring aromatics makes hydropyrolysis a potentially attractive route to chemicals from coal. The changes in the network of
thermal reactions engendered by the presence of hydrogen can be roughly classified as follows:

(i) During the early stages of pyrolysis, characterized by rapid tar release,
hydrogen penetrates the coal particle and reacts with various free radicals in
the gas phase or the condensed phase resulting in increased volatiles production.

(ii) The tar vapors react with hydrogen outside of the particles producing aromatic compounds of smaller molecular weight and, eventually, methane. These reactions include the degradation of condensed rings to single rings and the elimination of phenolic hydroxyl and alkyl substituents.

(iii) After the prolific formation of tar and gases has ceased, hydrogen reacts
with active sites on the residual char to produce methane. Initially rapid, this
reaction slows down considerably as the char is thermally deactivated.

Processes (ii) and (iii) correspond to what is normally called *hydropyrolysis*
or *flash hydrogenation* or *hydrocarbonization*. Sometimes (e.g. ref. 69) a distinction is made between hydropyrolysis, referring to relatively high temperatures
($600-1000^{o}C$) and short residence times, and hydrocarbonization referring to lower
temperatures ($450-600^{o}C$) and correspondingly longer residence times. Reactions
(iii), on the other hand, are characterized by the term *hydrogasification* because
they lead to a single product, methane.

In this chapter we will be concerned with hydropyrolysis, i.e. reaction groups
(i) and (ii) at the exclusion of hydrogasification which is more properly discussed in the context of coal gasification. In sections 7.1-7.4 we examine four
types of experimental systems for hydropyrolysis, the captive sample system, the
packed bed system, the modified captive sample system, and the entrained flow
system. Section 7.5 contains a review of model compound studies that relate mechanistically to coal hydropyrolysis. The final section 7.6 reviews kinetic modeling of hydropyrolysis.

7.1 CAPTIVE SAMPLE EXPERIMENTS

The apparatus and procedure used in these experiments are the same as the ones
used for straight pyrolysis (Section 4.1.2). In fact, the measurements reviewed
below were part of the pyrolysis program carried out by the MIT group. The captive sample technique allows relatively rapid removal of volatiles from the

reaction zone so that reactions in group (ii) are largely suppressed and hydropyrolysis is essentially limited to reaction group (i). This constitutes a limi-
tation of the captive sample technique from the standpoint of process-oriented
research, where reactions (ii) are utilized to produce the highly desirable single
ring aromatics.

Comparisons between weight loss- or total volatiles - under conditions of pyrolysis and hydropyrolysis have been made by Anthony et al. (ref. 125) and Suuberg
(ref. 63). Figures 5.8, 5.9 and 5.13 in Chapter 5 taken from the work of Anthony
et al. show the weight loss as a function of temperature, pressure and particle
size respectively. In Fig. **5.8**, the weight loss under 1 atm He (or N_2), 69 atm
He and 69 atm H_2 is the same until about 600°C above which the weight loss under
69 atm H_2 exceeds that under 1 atm He which in turn exeeds the weight loss under
69 atm He. In these experiments the sample was rapidly heated to its final temperature at which it was maintained for 5 to 20s.

Figure 7.1 shows the results that Suuberg obtained for the same bituminous
coal using a temperature time history consisting of a sharp pulse (see Section
5.2). The weight loss for all three atmospheres is identical, within experimental error, until a peak temperature of about 750°C. Above this temperature, the

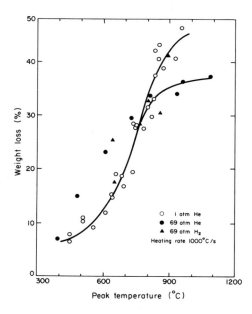

Fig. 7.1. Weight loss vs. peak temperature for
pyrolysis and hydropyrolysis of a bituminous coal
"Pittsburgh No. 8 (source: ref. 63).

140

weight loss is essentially the same at 69 atm H_2 and 1 atm He and exceeds that at 69 atm He.

In Figures 5.8 and 7.1, the weight loss curves start diverging at a temperature which marks the transition from conditions free of mass transfer limitations to conditions limited by mass transfer. The different transition temperatures, 600°C in Fig. 5.3 versus 750°C in Fig. 7.1 are evidently due to the different temperature time histories. Increasing the residence time at the highest tempera- ture, lowers the temperature of transition to mass transfer limitations.

Another effect of the prolonged residence time employed in the experiments of Fig. 5.3 is the higher weight loss at 69 atm H_2, compared to that at 1 atm He, a behavior which is not displayed for the pulse-like temperature histories of Fig. 7.1. This result can be attributed to the contribution of hydrogasification reactions (group iii) which is substantial only at the longer residence times. The increased weight loss at the longer residence times due to hydrogasification reactions is also evident in Figs. 5.10 and 5.13.

The most detailed measurements comparing product distributions in pyrolysis and hydropyrolysis were made by Suuberg et al. (refs. 63, 141). Figures 7.2 - 7.4 summarize some of their results.

Figure 7.2 compares the tar yields at 1 atm He, 69 atm He and 69 atm H_2. As we have already seen in Chapter 5 (Fig. 5.11), the yield at 1 and 69 atm He re- main the same until 700°C beyond which the yield at 69 atm drops considerably below the atmospheric yield. The yield at 69 atm H_2 is subject to competing effects. On

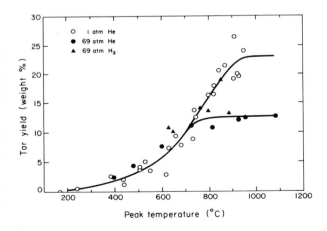

Fig. 7.2. Tar yield vs. peak temperature for pyrolysis and hydropyrolysis of a bituminous coal "Pittsburgh No. 8" (source: ref. 63).

the one hand, hydrogen stabilizes free radicals susceptible to reattachment in the condensed phase. On the other hand, hydrogen reacts with tar precursors in the voids or in the coal melt to produce lower molecular weight products and, at the same time, the increased pressure suppresses the rate of mass transfer away from the particle. The scatter in the data of Fig. 7.2 does not allow quantitative assessment as to the relative magnitude of these effects.

Figure 7.3 compares the yields of methane in 1 atm He and 69 atm H_2. At all temperatures the yield in hydrogen considerably exceeds the yield in the low pressure inert environment. The large differences in the yield are evidently due to the synergism of the two factors mentioned earlier. High pressure reduces the rate of mass transfer and thus increases the probability of secondary reactions including reactions of hydrogenolysis of tar vapors. Additional contributors to the increased methane yield are reaction of molecular hydrogen with active sites in the coal matrix that are not associated with tar precursors. Such reactions include the elimination of methyl substituents on aromatic rings. In connection with Figs. 7.1 - 7.3 it must be noted that the effects of hydrogen on tar and gases are in the opposite direction, whence the more modest effect on total weight loss.

In addition to the bituminous coal, Suuberg studied a lignite with respect to product yields under conditions of pyrolysis and hydropyrolysis (refs. 63, 141). As shown on Fig. 7.4, starting with about 500°C, the methane yield under 69 atm H_2 exceeds the yields under 1 and 69 atm He. The latter two yields remain equal until about 700°C which marks the inception of mass transfer limitations.

Increased yields of hydrocarbon gases other than methane and ethylene were similarly observed in the presence of hydrogen at temperatures as low as 500°C. The low temperature marking the deviation between the gas yields from pyrolysis and hydropyrolysis signifies as before that hydrogen does not only influence the course of secondary reactions of tar precursors but participates in direct reactions with the coal matrix.

In contrast to the yields of other hydrocarbon gases, the yields of ethylene at 1 atm He and 69 atm H_2 were equal and, beginning at 700°C, surpassed the yield at 69 atm He.

Although the amount of tar obtained from lignite was low and, hence, subject to larger measurement error, it could be still observed that the tar yields at 1 atm He and 69 atm H_2 were higher than the yield at 69 atm He. The weight loss at 69 atm H_2 slightly exceeded that at 1 atm He. Compared to the bituminous coal, lignite displays a somewhat different weight loss dependence on total pressure and hydrogen pressure probably due to the difference in the *relative* tar yields between the two coals.

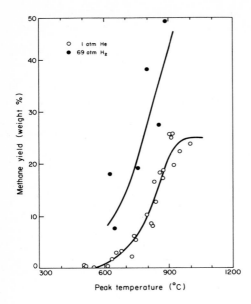

Fig. 7.3. Methane yield vs peak temperature
for pyrolysis and hydropyrolysis of a bituminous
coal "Pittsburgh No. 8" (source: ref. 63).

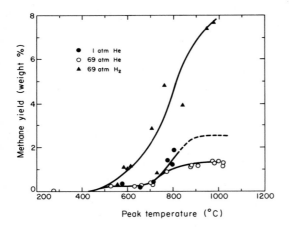

Fig. 7.4. Methane yield vs. peak temperature for pyrolysis
and hydropyrolysis of a lignite (source: ref. 63).

7.2 PACKED BED EXPERIMENTS

In this arrangement, the coal sample is held stationary in a section of a
tubular reactor, which we shall call the *hydropyrolysis section*, where it is subjected to a temperature program under hydrogen flow.   The volatile products carried in the hydrogen stream pass through an additional heated section, which we
shall call the *hydrogenolysis section*, and after quenching are conducted to product collection and sampling equipment. By regulating the hydrogen flow rate
and suitably controlling the temperatures in the hydropyrolysis and hydrogenolysis sections of the tubular reactor it is possible, in principle, to control the
temperature and residence time of the solid and the volatile products independently.

In early experiments by Hiteshue et al. (refs. 142, 143) utilizing the packed
bed arrangement, the heating period was relatively long, about three minutes, and
the temperatures of the solid sample and the volatile products could not be controlled independently. In a recent study, Finn et al. (ref. 144) used a two-segment tube to implement independent temperature control of solids and volatiles
while achieving heating times as short as half a minute.

A schematic of the apparatus used by Finn et al. is shown in Fig. 7.5. Coal
was placed in an 8 cm long bed in the hydropyrolysis section and subjected to a

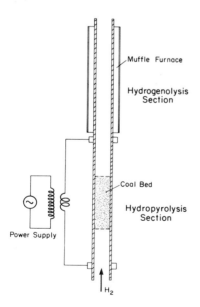

Fig. 7.5. Schematic of two-segment hydropyrolysis
reactor used by Finn et al. (ref. 144).

temperature pulse by direct resistive heating of the tube wall using a low volt-
age transformer. The hydrogenolysis section was maintained at constant temperature by a muffle furnace. The reactor tube was 8 mm ID and the temperature was
recorded at the tube wall. One disadvantage of using a massive coal sample was
that the true heating period was probably considerably longer than the half minute reported for the tube wall. Another disadvantage was the extensive secondary
reactions of tar vapors and other volatiles on the coal surface before entering
the second section intended for hydrogenolysis.

Figures 7.6 - 7.9 show some of the results of Finn et al. (ref. 144). The
yield of various single ring aromatics vs. peak temperature is shown in Fig. 7.6.
The temperature pulse, common in both reactor sections, consisted of a rising
segment (heating rate 7^OK/s) immediately followed by rapid cooling (three seconds).
The products consisted of approximately equal amounts of benzene-toluene-xylene
(BTX) and phenol-cresols-xylenols (PCX). Both classes of products passed through
a maximum at a temperature slightly below 1000^OK. The maximum yield of BTX + PCX
was about 5 per cent.

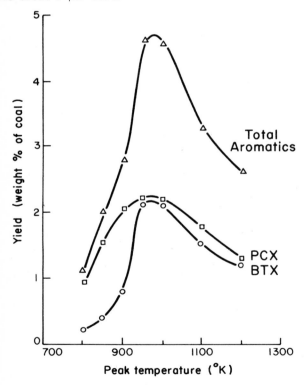

Fig. 7.6. Yield of single ring aromatic products
vs. peak temperature for hydropyrolysis of a bituminous coal at 150 bar pressure and 11s hydrogenolysis
residence time (source: ref. 144).

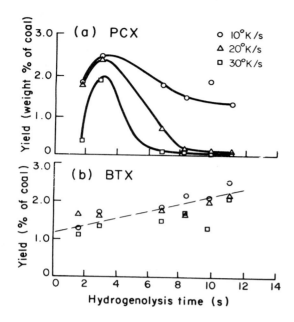

Fig. 7.7. Yield of BTX and PCX vs. hydrogenolysis time at different heating rates for
hydropyrolysis of a bituminous coal at 150 bar
pressure and 1000°K peak temperature (source:
ref. 144).

Fig. 7.7 plots the product yields obtained under the same type of temperature
pulse as in Fig. 7.6 but for different heating rates and hydrogenolysis times.
The yield of BTX increases steadily with hydrogenolysis time and is rather insen-
sitive to heating rate. These trends can be explained by the fact that BTX is
an intermediate product resulting from the degradation of tar and in turn being
converted to methane. Since the latter reaction is slower, the maximum of BTX
corresponds to hydrogenolysis times larger than ten seconds and is not shown in
the figure. The effect of heating rate is smaller and largely masked by the
scatter in the data.

In contrast to the yield of BTX, the yield of PCX shows some rather striking
trends. At fixed heating rate, the yield passes through a maximum at about three
seconds hydrogenolysis time. The presence of this maximum suggests consecutive
reactions from tars to PCX and PCX to BTX or directly to methane. Since the
decrease in PCX is not accompanied by a commensurate increase in BTX, the direct
conversion of PCX to methane seems to be the predominant route. At fixed

hydrogenolysis time, PCX decreases rather rapidly with increasing heating rate probably due to the shorter *solids* exposure to high temperatures decreasing the yield of precursor tar vapors. Why this same effect is not shown by the yield of BTX remains a vexing question.

A different temperature-time program was used in the measurements reported in Figs. 7.8, 7.9. After rising to its maximum value, the temperature in the hydropyrolysis section was maintained constant for ten to fifteen minutes while the temperature in the hydrogenolysis section was kept at some other constant value throughout the run Figure 7.8 shows various product yields vs. hydrogen-olysis temperature. Since the hydrogen flux was kept constant in these runs, variation of the hydrogenolysis temperature was accompanied by variation of the hydrogenolysis time. Nevertheless, the yield curves still reflect the fact that tar vapors are precursors for benzene and other light aromatics which in turn are converted to the final product methane. Figure 7.9 plots the yields of several products vs. hydropyrolysis temperature. The maximum yield of benzene, about 12 percent, is quite promising from the standpoint of producing chemicals from coal.

7.3 MODIFIED CAPTIVE SAMPLE EXPERIMENTS

To achieve high heating rates and prevent secondary reactions on the coal par-ticle surface, Graff et al. (refs. 145, 146) developed an experimental technique

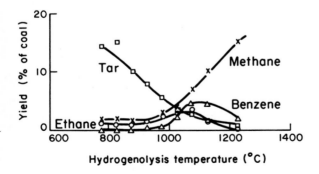

Fig. 7.8. Product yields vs. hydrogenolysis temperature for a bituminous coal at heating rate 1°K/s, peak hydro-pyrolysis temperature 750°K and hydrogen pressure 100 bar (source: ref. 144).

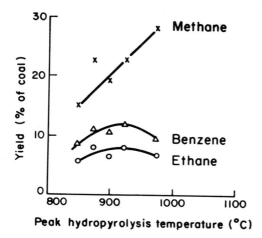

Fig. 7.9. Product yields vs. peak hydropyrolysis temperature for a bituminous coal at heating rate 5°K/s, hydrogenolysis temperature 1123°K and hydrogen pressure 150 bar
(source: ref. 144).

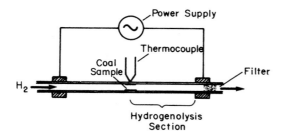

Fig. 7.10. Modified captive sample reactor for coal
hydropyrolysis (source: ref. 145).

combining the advantages of the captive sample and the packed bed techniques.
The reaction section of their setup is shown in Fig. 7.10. The reactor consists
of a stainless steel tube 5.1 mm ID, 6.3 mm OD and 30 cm length capable of with-
standing up to 1000°C temperature and 100 atm pressure. The finely ground coal
is deposited on a circular region in the middle of the tube. The whole reactor
tube is heated resistively by means of a DC power supply switched on and off by
a control circuit. As with the captive sample equiment described in Chapter 4,
resistive heating is applied at two levels. The first and higher level serves
to heat the tube to the preset temperature at a rate up to 1500°C/s. After the
desired temperature is established, a control circuit switches power to the lower

level, adequate to maintain the tube at the desired steady temperature for the
duration of the experiment. A spot welded thermocouple serves to indicate the
temperature and activate the switching circuit.

After establishing the hydrogen flow at the desired pressure and flow rate,
the power supply is switched on and the volatiles released from the coal sample
are carried in the hydrogen stream through the downstream section of the tube
which constitutes the section for hydrogenolysis. While this experimental setup
provides identical hydropyrolysis and hydrogenolysis temperatures, the sample
heating technique can be applied in conjunction with a two-segment reactor with
separate control of the hydrogenolysis temperature. The residence time in the
hydrogenolysis section is in all cases controlled by the hydrogen mass flow rate,
with due allowance for the volumetric expansion at the reaction temperature.

The reaction products are collected in evacuated tanks from which samples are
drawn for analysis by gas chromatography. An ingenious technique is used to
prevent undue dilution by hydrogen of the product gas. A thermal conductivity
cell detects the level of products in the product stream and only when this level
is above a preset value is the product stream directed to the sample tanks. The
residual char is determined by oxidation in place and analysis of the carbon
oxides produced. Heavy liquids not detectable by gas chromatography, are reported as "carbon deficit."

Some of the results reported by Graff et al. (refs. 145, 146) for a high volatile bituminous coal (Illinois No. 6) are reproduced in Figs. 7.11, 7.12 with
the yields expressed as carbon in the products as a percentage of carbon in the
coal. Figure 7.11 shows the yields of methane, ethane and propane vs. reaction
temperature for fixed solids and vapors residence time. The monotonically increasing yield of methane is obviously due to the fact that this gas is the final
hydrogenolysis product of tar vapors and hydrocarbon gases. Ethane, on the other
hand as an intermediate product passes through a maximum. The monotonic decrease
of propane might be due to the decomposition of its precursor propyl radicals to
ethylene and methyl radicals, favored at the higher temperatures.

Figure 7.12 plots the yields of BTX and tar versus reaction temperature. The
tar was determined indirectly as the difference between the original carbon and
the carbon in all measured products, including char. The determination by difference is obviously subject to considerable error. As in the studies discussed in
Section 7.2, the yield of BTX passes through a maximum of about 12 percent. This
yield is comparable to that shown in Fig. 7.8 corresponding to much lower heating
rates. The temperature of the maximum was in both cases about 800°C. The simi-
larity of the results in Figs. 7.8 and 7.12 indicates that, isolated from other
operating variables, heating rate has a relatively minor effect on product yields.

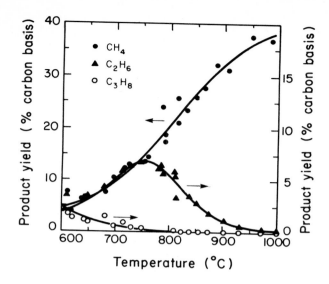

Fig. 7.11 Yield of gaseous hydrocarbons vs. temperature for hydropyrolysis of a bituminous coal "Illinois No. 6" at 100 atm H_2 and 0.6s hydrogenolysis
time (source: ref. 146).

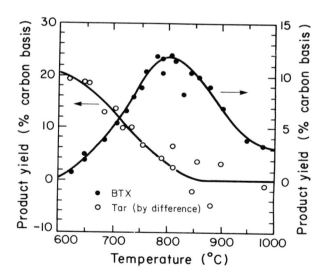

Fig. 7.12. Yields of BTX and tar (by difference) vs. temperature for hydropyrolysis of a
bituminous coal "Illinois No. 6" at 100 atm
H_2, 650°C/s heating rate and 0.6s hydrogenolysis time (source: ref. 146).

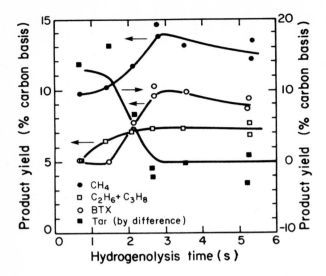

Fig. 7.13. Product yields vs. hydrogenolysis time for hydropyrolysis of a bituminous coal "Illinois No. 6" at 700°C, 100 atm H_2 and 650°C/s heating time (source: ref. 146).

Figure 7.13 shows the dependence of product yields on hydrogenolysis time (volatiles residence time) at 700°C. The maximum in the BTX yield is attained at about three seconds. This optimal time would decrease to a fraction of a second at 800°C as inferred from the previous figure. The shallow maximum in the yield of methane is somewhat puzzling in view of the slow decline of BTX at times larger than three seconds.

7.4 ENTRAINED FLOW EXPERIMENTS

While the modified captive sample reactor is well suited to fundamental kinetic studies, the entrained flow reactor is better suited to process development by being amenable to scale-up to pilot plant units. Entrained flow experiments are difficult to interpret kinetically because coal particles react continuously during their passage through the reactor, therefore, the volatile products have a distribution of residence times for hydrogenolysis. On the other hand, the entrained flow reactor operating at steady state generates large samples suitable for the accurate measurement of product yields. An entrained flow reactor for hydropyrolysis generally has geometry similar to an entrained flow pyrolysis reactor, but must be capable of operating at high pressures.

An entrained flow system was used by Fallon et al. (ref. 147) to study the hydropyrolysis of a lignite and a subbituminous coal at 700-900°C and 500-1,500 psia of hydrogen. The products up to and including BTX were determined by gas chromatography while the heavier liquids were collected and analyzed at the end

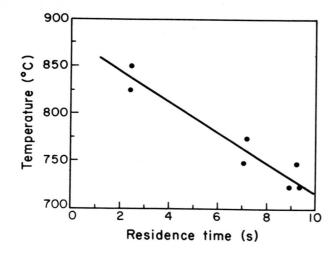

Fig. 7.14. Locus of maximum BTX yield for hydropyrolysis
of lignite at 2500 psia H_2 (source: ref. 147).

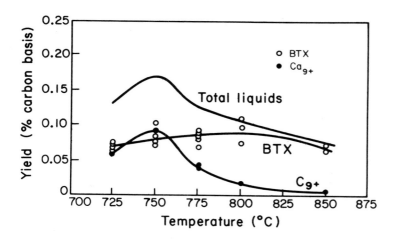

Fig. 7.15. Maximum yields of BTX and C_{9+} liquids vs. temperature for hydropyrolysis of lignite at 2000 psia H_2
(source: ref. 147).

of each experiment. Several runs with lignite explored the effect of the three
principal variables on the yield of products, especially BTX. At fixed residence time and pressure, the yield of BTX passes through a maximum in the range
700-800°C. Likewise, at fixed temperature and pressure, the yield of BTX becomes
maximum at some intermediate residence time, past which it declines rapidly to

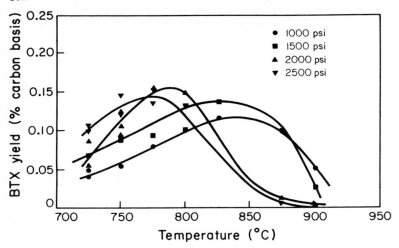

Fig. 7.16. Maximum BTX yield vs. temperature for hydropyrolysis of a subbituminous coal at various hydrogen pressures (source: ref. 147).

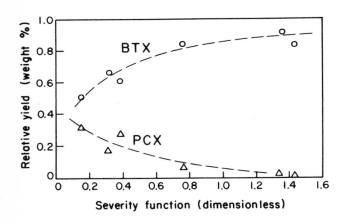

Fig. 7.17. Relative yields of BTX and PCX in the liquid products as a function of severity for the hydropyrolysis of a lignite at 2000 psia H_2 (source: ref. 148).

zero. On the other hand, at fixed residence time, the yield vs. temperature curve is rather broad around the maximum value. The maximum yields under most conditions were in the range 8-10 percent in terms of carbon conversion. Figure 7.14 is a locus of temperature-residence time conditions under which the BTX yield was near its maximum value of 8-10 percent. The maximum yields are shown in Fig. 7.14

to increase as the hydrogen pressure increased from 500 to 2000 psia. Upon further increasing the pressure to 2500 psia, the maximum yield remained essentially unchanged but the residence time required to attain this yield decreased.

Liquids heavier than BTX (C_{9+}) were also observed in significant yields as shown in Fig. 7.15. These liquids were much lighter than pyrolysis tars consisting of about 40 percent naphthalene and only trace amounts of phenols. However, being considerably more reactive than BTX, they readily declined with increasing temperature.

Hydropyrolysis experiments performed using a subbituminous coal resulted in BTX yields as high as 15 percent compared to the 10 percent obtained with lignite. Figure 7.16 shows the maximized yield of BTX (with respect to residence time) as a function of temperature at four pressure levels.

In another recent study, Beeson et al. (ref. 148) studied the hydropyrolysis of a lignite using an entrained flow reactor with controlled axial temperature profiles. Although the intent was to determine the effect of the heating rate, the ability to vary the axial temperature profile offers a potentially useful variable for product optimization.

The results reported are particularly interesting relative to the detailed breakdown of liquids in the gasoline boiling range into several fractions: BTX, C_{9+} aromatics, indenes + indans, phenols + cresols and naphthalene. Overall yields of these liquid products (carbon in the liquids as a fraction of carbon in the original coal) ranged between 0.07 and 0.15. The mass fraction of phenols and cresols in the liquids was as high as 0.76 indicating that the phenolic compounds constitute the primary hydropyrolysis products from lignite. The phenolic products react further to BTX and methane at a rate depending on temperature and hydrogen pressure.

Figure 7.17 shows the variation of the relative yields of BTX and phenol + cresol as a function of a severity parameter defined by

$$severity = \int_0^t k_0 \, dt$$

where $k_0 = 9 \times 10^5 \exp(-30,700/RT)$. The rate constant k_0 was assigned by reference to some earlier data on anthracene hydrogasification, therefore, the severity parameter is a somewhat arbitrary measure of the combined effect of temperature and residence time.

Two other related hydropyrolysis programs with emphasis on process and hardware development are the Cities Service short residence time hydropyrolysis program (ref. 149) and the Rockedyne program (ref. 150). The City Service program has employed a laboratory scale entrained flow reactor (about 1 Kg coal/hr) while the Rockedyne program has utilized a process development entrained flow reactor (about 200 Kg coal/hr). The distinguishing feature of the second reactor is the

rapid mixing between feed coal and hydrogen, achieved by a "rocket engine" injector. Results reported to date for a lignite showed maximum BTX yields of about
10 percent (City Service) or total liquid yields of about 30-40 percent (Rockedyne).

7.5 MODEL COMPOUND STUDIES

As mentioned at the beginning of the chapter, hydropyrolysis reactions include
(i) reactions of hydrogen with the condensed phase during the stage of liquids
formation (ii) hydrogenolysis of the vapors in the gas phase to produce PCX
(phenols), BTX and light hydrocarbon gases. The model compound studies discussed
below are useful primarily in understanding the mechanism and kinetics of hydropyrolysis reactions in class (ii) which we have earlier labelled as hydrogenolysis. Some general issues that are of particular interest are the mechanisms of
degradation of ring systems, e.g. naphthalene to toluene or benzene to methane;
and the mechanisms of dealkylation and dehydroxylation, e.g. toluene to benzene
or cresol to toluene. In addition to reaction pathways and mechanisms, it would
be valuable to possess a reasonable kinetic description of the effect of operating variables on product yields.

Virk et al. (ref. 151) analyzed existing data on unsubstituted aromatic hydrocarbons and found that the rates of disappearance of each compound in pyrolysis
and hydrogenolysis were roughly equal, although the products were different. In
the absence of hydrogen, successive condensation and dehydrogenation led to a
final solid product, coke. In the presence of hydrogen, the final product was
methane. Intermediate products with a smaller number of fused rings were not
specifically identified. Based on the approximate equality of the rates of pyrolysis and hydrogenolysis they proposed that both reactions have a common rate determining step, namely the "destabilization" of the aromatic ring. Although the
mechanism of this step was not identified, its rate was assumed to be related to
the ring delocalization energy.

Penninger and Slotboom (ref. 152) reviewed experimental data on the hydrogenolysis of several substituted and unsubstituted aromatics. For the case of unsubstituted naphthalene and phenanthrene they concluded that ring cracking occurs
through the formation of an intermediate hydroaromatic compound. For example,
naphthalene is first hydrogenated to tetralin which subsequently decomposes to
various alkylbenzenes. The mechanism of the crucial first step, the hydrogenation
of the unsubstituted aromatic, was not identified. On the other hand, the subsequent hydrogenolysis of the hydroaromatics was explained by free radical
mechanisms.

The hydrogenolysis of hydroaromatics can be illustrated with the reactions of
tetralin which have already been discussed in a different context (Section 6.3.2).
We are here interested in the mechanism of utilization of molecular hydrogen.
One possibility is offered by the reaction

$$R\cdot + H_2 \rightleftharpoons H\cdot + RH$$

where $R\cdot$ is a carbon centered radical, more specifically an alpha radical. Despite the unfavorable equilibrium (ΔG is about 18,800 at $300^{\circ}K$), this reaction increases the concentration of hydrogen atoms which can then participate in addition reactions such as

The dihydronaphthalene produced can be subsequently hydrogenated to naphthalene by the same mechanism, via the addition of a hydrogen atom. No mechanism has so far been proposed for the direct (pericyclic) addition of molecular hydrogen, although the possibility cannot be excluded.

Increased concentration of hydrogen atoms due to the presence of molecular hydrogen is effective in tetralin decomposition via addition and opening of the saturated ring (Section 6.3.2) and in dealkylation. The latter reaction may be illustrated by the example

with the methyl radical ending up as methane after hydrogen abstraction.

Cypres and Bettens (refs. 47-49) studied the pyrolysis of phenol and cresols in the absence of hydrogen and proposed non-free-radical mechanisms for these reactions (see Section 3.7). The mechanisms proposed leave some open questions and cannot be readily extended to include the effect of molecular hydrogen.

7.6 MODELING

We start by recalling the classification of reactions into groups (i)-(iii) defined at the beginning of the chapter. Most of the models concerning hydrogen-coal reactions have been addressed to reaction group (iii) in the context of coal gasification to methane. Models of this type will not be discussed here since they are not relevant to the early phases of hydropyrolysis. Reaction groups (i) and (ii), dominating the early phases of hydropyrolysis, have been considered in only a few modeling studies, three of which are discussed below.

The experimental work of Anthony et al. (refs. 125, 126) and Suuberg et al. (refs. 63, 141) have demonstrated the effects of inert pressure, hydrogen pressure and particle size on the total yield of volatiles (see e.g. Figs. 5.4, 5.9, 7.1). To quantitatively describe such effects which are intimately related to mass transfer limitations Anthony et al. (ref. 125) proposed the following set of phenomenological reactions

$$\text{coal} \rightarrow \nu_1 V^* + \nu_2 V + S^* \tag{7.1}$$

$$V^* + H_2 \rightarrow V \tag{7.2}$$

$$V^* \rightarrow S \tag{7.3}$$

$$S^* + H_2 \rightarrow V + S^* \tag{7.4}$$

$$S^* \rightarrow S \tag{7.5}$$

Coal decomposes to "unreactive" volatiles V, reactive volatiles V^* and a reactive solid S^*. The unreactive volatiles consist of gases such as methane, steam, carbon oxides, light liquids (e.g. BTX) and heavier products, tar. The reactive volatiles presumably consist of free radicals or other unstable molecules. S^* is a reactive solid susceptible to hydrogenation by the fourth reaction while S is a solid which participates in no further reactions in the time scale of interest.

Mass transfer enters in the problem through a balance for species V^* in the voids of the coal particle. Assuming steady state conditions, the mass balance becomes

$$K(c^* - c_\infty^*) = \nu_1 r_1 - r_2 - r_3 \tag{7.6}$$

where c^*, c_∞^* are the concentrations of V^* inside and outside the particle, r_1, r_2,... are the rates of reactions (7.1), (7.2),...., and K is a mass transfer coefficient. The reaction rates were expressed in first or second order form and the rate constants were specified numerically to match the experimental data (ref. 125). From our standpoint it is important to take notice of the fundamental assumptions or approximations of the model which were (i) the gas space inside the particle has fixed volume and uniform composition (ii) the concentration of hydrogen inside the particle is uniform and equal to the outside concentration (iii) all reactive species can be lumped into one, V^* (iv) the mass transfer coefficient is inversely proportional to the pressure but independent of particle size. The yield of total volatiles calculated on the basis of this model was in most respects in good agreement with the experimental yield. However, the model predicted a stronger pressure dependence than experimentally observed while it failed to predict the observed effect of particle size in the presence of hydrogen. Both points of disagreement seem to derive from assumption (ii) and to suggest the existence of a hydrogen pressure drop from the outside to the inside of the particle.

In a relatively recent study (ref. 135), Russel et al. carried out an elegant and comprehensive theoretical analysis of hydropyrolysis reactions coupled with intraparticle mass transfer. They employed a reaction system identical to (7.1)-(7.5) except that reaction (7.2) was assumed instantaneous, therefore V^* and H_2 disappeared on a reaction front gradually progressing towards the center of the particle. Mass transfer was described by the "dusty gas" model, taking into account fluxes due to diffusion and pressure gradients. For this purpose the coal particle was assumed to possess a stable pore structure, an assumption which applies reasonably well to nonsoftening coals but not to sofening coals (see Chapter 5). The model was nonetheless tested against the pyrolysis and hydro-pyrolysis data from a high volatile bituminous coal (refs. 125, 126).

Recent hydropyrolysis modeling work by the MIT group was presented by Schaub et al. (ref. 153). Although this publication gives very few details, the two basic premises of the analysis can be summarized as follows:

(i) The hydropyrolysis reactions are represented by the scheme below where M is the familiar by now metaplast (Chapter 5), S is a reactive solid termed "semi-coke" and A is another reactive intermediate in the condensed phase. Step 3 is the transport of metaplast molecules from the coal melt to the gas phase as tar. All other steps are chemical in nature.

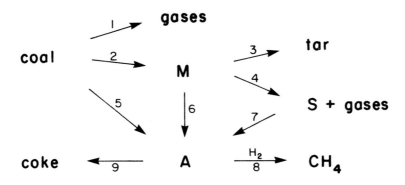

(ii) Step 8 requires the diffusion of hydrogen in the condensed phase which is initially a melt but later becomes a solid, char.

The reaction scheme shown above differs considerably from the one utilized in refs. 125, 135. It employs a more complex network of consecutive reactions and differentiates between three products, gases, methane and tar. It also assumes that hydrogen reacts with a species in the condensed phase rather than the gas phase and predicts that the tar yield depends on total pressure but not on the nature of the surrounding gas. Unfortunately, the limited experimental data available (Fig. 7.2) are insufficient to test this crucial prediction. Another noteworthy feature of the model is the consideration of hydrogen diffusion through the coal melt, enhanced by the stirring action of the evolving bubbles. Although some of its detailed assumptions could be questioned, this model represents the most physically realistic effort in hydropyrolysis modeling.

REFERENCES

1   J. G. Speight, Appl. Spec. Rev., 5 (1971) 211-264.
2   P. H. Given, Lecture Notes, 1979.
3   H. L. Retcofsky, T. J. Brendel and R. A. Friedel, Nature, 240 (1972) 18-19.
4   D. D. Whitehurst, in J. N. Larsen (ed.), Organic Chemistry of Coal, ACS
5   R. R. Ruberto, D. C. Cronauer, D. M. Jewell and K. S. Seshadri, Fuel, 56
    (1977) 17-24.
6   G. L. Tingey and J. R. Morrey, Battelle Energy Program Report, Battelle,
    Pacific Northwest Laboratories, 1972.
7   G. R. Gavalas and M. Oka, Fuel, 57 (1978) 285-288.
8   K. D. Bartle, T. G. Martin and D. F. Williams, Fuel, 54 (1975) 226-235.
9   K. D. Bartle, W. R. Ladner, T. G. Martin, C. E. Snape and D. F. Williams,
    Fuel, 58 (1979) 413-422.
10 S. Yokohama, D. M. Bodily and W. H. Wiser, Fuel, 58 (1979) 162-170.
11 C. E. Snape, W. R. Ladner and K. D. Bartle, Anal. Chem., 51 (1979) 2189-2198.
12 N. C. Deno, B. A. Greigger and S. G. Stroud, Fuel, 57 (1978) 455-459.
13 N. C. Deno, K. W. Curry, B. A. Greigger, A. D. Jones, W. G. Rakitsky, K. A.
    Smith, K. Wagner and R. D. Minard, Fuel, 59 (1980) 694-698.
14 N. C. Deno, B. A. Greigger, A. D. Jones, W. G. Rakitsky, K. A. Smith and
    R. D. Minard, Fuel, 59 (1980) 699-700.
15 J. D. Brooks, R. A. Durie and S. Sternhell, Aust. J. Appl. Sci., 9 (1958)
    63-80.
16 G. Alberts, L. Lenart, H. H. Oelert, Fuel, 53 (1974) 47-53.
17 S. Friedman, M. L. Kaufman, W. A. Steiner and I. Wender, Fuel, 39 (1960) 33-
    45.
18 B. S. Ignasiak and M. Gawlak, Fuel, 56 (1977) 216-222.
19 D. C. Cronauer and R. G. Ruberto, Electric Power Research Institute Report
    AF-913, 1979.
20 J. K. Brown, W. R. Ladner and N. Sheppard, Fuel, 39 (1959) 79-86.
21 J. K. Brown and W. R. Ladner, Fuel, 40 (1960) 87-96.
22 H. L. Retcofsky, F. K. Schweighardt and M. Hough, Anal. Chem., 49 (1977)
    585-588.
23 K. D. Bartle and J. A. S. Smith, Fuel, 44 (1965) 109-124.
24 A. A. Herod, W. R. Ladner and C. E. Snape, Prepr. Royal Society Meeting on
    New Coal Chemistry, Stoke Orchard, 1980.
25 E. M. Dickinson, Fuel, 59 (1980) 290-294.
26 K. D. Bartle and D. W. Jones, in C. Karr, Jr. (Ed.), Analytical Methods for
    Coal and Coal Products, Vol. II, Academic Press, New York, 1978, pp. 103-160.
27 M. Oka, H. C. Chang and G. R. Gavalas, Fuel, 56 (1977) 3-8.
28 M. Zander and J. W. Stadelhofer, private communication.
29 S. W. Benson and H. E. O'Neal, Kinetic Data on Gas Phase Unimolecular Reac-
    tions, NSRDS-NBS 21, 1970.
30 S. W. Benson, Thermochemical Kinetics, Wiley, New York, 1976.
31 S. E. Stein and D. M. Golden, J. Org. Chem., 42 (1977) 839-841.
32 B. D. Barton and S. E. Stein, J. Phys. Chem., (1980) 2141-2145.
33 B. W. Jones and M. B. Neuworth, Ind. Eng. Chem., 44 (1952) 2872-2876.
34 K. J. Laidler, Chemical Kinetics, McGraw-Hill, New York, 1965, pp. 199.
35 D. A. Robaugh, R. E. Miller and S. E. Stein, Prepr. Div. Fuel Chem., Am.
36 A. M. North and S. W. Benson, J. Am. Chem. Soc., 84 (1962) 935-940.
37 R. D. Burkhart, J. Pol. Sci. A, 3 (1965) 883-894.
38 J. N. Cardenas and K. F. O'Driscoll, J. Pol. Sci. A, 14 (1976) 883-897.
39 D. F. McMillen, W. C. Ogier and D. S. Ross, Prepr. Div. Fuel Chem., Am. Chem.
    Soc., 26 No. 1 (1981) 181-190.
40 A. A. Zavitsas and A. A. Melikian, J. Am. Chem. Soc., 97 (1975) 2757-2763.
41 J. A. Kerr and M. J. Parsonage, Evaluated Kinetic Data on Gas Phase Addition
    Reactions, Butterworths, London, 1972.
42 M. Levy and M. Szwarc, J. Am. Chem. Soc., 77 (1955) 1949-1955.
43 M. Szwarc and J. H. Bink, in Theoretical Organic Chemistry, Kekulé Symposium,
    Butterworths Scientific, London, 1959, pp. 262-290.

160

44   G. H. Williams, Homolytic Aromatic Substitution, Pergamon Press, New York, 1960.
45   L. W. Vernon, Fuel, 59 (1980) 102-106.
46   J. D. Brooks, R. A. Durie and S. Sternhell, Aust. J. Appl. Sci. 9, (1958) 303-320.
47   R. Cypres and B. Bettens, Tetrahedron, 30 (1974) 1253-1260.
48   R. Cypres and B. Bettens, Tetrahedron, 31 (1975) 353-357.
49   R. Cypres and B. Bettens, Tetrahedron, 31 (1975) 359-365.
50   P. S. Virk, D. H. Bass, C. P. Eppig and D. J. Ekpenyong, in Whitehurst (ed) Coal Liquefaction Fundamentals, ACS Symposium Series No. 139, Am. Chem. Soc., New York, 1980, pp. 329-345.
51   W. von E. Doering and J. W. Rosenthal, J. Am. Chem. Soc., 89 (1967) 4534-4535.
52   B. M. Benjamin, V. F. Raanen, P. H. Maupin, L. L. Brown and C. J. Collins, Fuel, 57 (1978) 269-272.
53   P. Bredael and T. H. Vinh, Fuel, 58 (1979) 211-214.
54   D. C. Cronauer, D. M. Jewell, Y. T. Shah and K. A. Kueser, Ind. Eng. Chem. Fundamentals, 17 (1978) 291-297.
55   R. T. Eddinger, L. D. Friedman and E. Rau, Fuel, 45 (1966) 245-252.
56   S. Badzioch and P. G. W. Hawksley, Ind. Eng. Chem. Process Design Develop. 9 (1970) 521-530.
57   H. Kobayashi, J. B. Howard and A. F. Sarofim, Sixteenth Symposium (International) on Combustion, The Combustion Institute, Pittsburgh, 1977, pp. 411-425.
58   S. K. Ubhayakar, D. B. Stickler, C. W. von Rosenberg, Jr. and R. E. Gannon, Sixteenth Symposium (International) on Combustion, The Combustion Institute, Pittsburgh, 1977, pp. 427-436.
59   P. R. Solomon, D. G. Hamblen, G. J. Goetz and N. Y. Nsakala, Prepr. Div. Fuel Chem., Am. Chem. Soc., 26 No. 3 (1981) 6-17.
60   D. B. Anthony, J. B. Howard, H. P. Meissner and H. C. Hottel, Rev. Sci. Instrum., 45 (1974) 992-995.
61   P. R. Solomon and M. B. Colket, Seventeenth Symposium (International) on Combustion, The Combustion Institute, Pittsburgh, 1979, pp. 131-143.
62   G. R. Gavalas and K. A. Wilks, A.I.Ch.E. J. , 26 (1980) 201-212.
63   E. M. Suuberg, Sc.D. Thesis, Massachusetts Institute of Technology, 1977.
64   E. M. Suuberg, W. A. Peters and J. B. Howard, Ind. Eng. Chem. Process Design Develop. 17 (1978) 34-46.
65   R. Jain, Ph.D. Thesis, California Institute of Technology, 1979.
66   H. Jüntgen and K. H. van Heek, paper presented to the meeting on coal fundamentals, Stoke Orchard, 1977.
67   D. W. Blair, J. O. L. Wendt and W. Bartok, Sixteenth Symposium (International) on Combustion, The Combustion Institute, Pittsburgh, 1977, pp. 475-489.
68   B. Mason Hughes and J. Troost, Prepr. Div. Fuel Chem., Am. Chem. Soc., 26 No. 2 (1981) 107-120.
69   J. B. Howard, W. A. Peters and M. A. Serio, EPRI AP-1803, 1981.
70   P. R. Solomon, United Technologies Research Center Report R77-952588-3, 1977.
71   E. M. Suuberg, W. A. Peters and J. B. Howard, Seventeenth Symposium (International) on Combustion, The Combustion Institute, Pittsburgh, 1979, pp. 117-130.
72   E. M. Suuberg, W. A. Peters and J. B. Howard, in Oblad (Ed), Thermal Hydrocarbon Chemistry, Advances in Chemistry Series, No. 183, Am. Chem. Soc., New York, 1979, pp. 239-257.
73   W. Peters and H. Bertling, Fuel, 44 (1965) 317-331.
74   H. Jüntgen and K. H. van Heek, Fuel, 47 (1968) 103-117.
75   R. Cypres and C. Soudan-Moinet, Fuel, 59 (1980) 48-54.
76   P. R. Solomon, United Technologies Research Center Report R76-952588-2, 1977.
77   J. H. Pohl and A. F. Sarofim, Sixteenth Symposium (International) on Combustion, The Combustion Institute, Pittsburgh, 1977, pp. 491-501.
78   P. R. Solomon, Prepr. Div. Fuel Chem., Am. Chem. Soc., 24 No. 2 (1979) 179-184.
79   P. R. Solomon and M. B. Colket, Fuel, 57 (1978) 749-755.

80 H. C. Howard, in H. H. Lowry (Ed.), Chemistry of Coal Utilization, Suppl. Volume, Wiley, New York, 1963, pp. 340-394.
81 A. H. Billington, Fuel, 33 (1954) 295-301.
82 A. J. Forney, R. F. Kenny, S. J. Gasior and J. H. Field, I & EC Product Res. Dev., 3 (1964) 48-53.
83 D. J. McCarthy, Fuel, 60 (1981) 205-209.
84 O. P. Mahajan, M. Komatsu and P. L. Walker, Jr., Fuel, 59 (1980) 3-10.
85 J. F. Jones, M. R. Schmid and R. T. Eddinger, Chem. Eng. Progr., 60 No. 6 (1964) 69-73.
86 W. H. Ode, in H. H. Lowry (Ed.), Chemistry of Coal Utilization, Suppl. Volume, Wiley, New York, 1963, pp. 202-231.
87 O. P. Mahajan, in R. A. Meyers (Ed.) Coal Structure, Academic Press, New York (in press).
88 H. N. S. Schafer, Fuel, 58 (1979) 673-679.
89 H. N. S. Schafer, Fuel, 59 (1980) 295-301.
90 H. N. S. Schafer, Fuel, 59 (1980) 302-304.
91 R. J. Tyler and H. N. S. Schafer, Fuel, 59 (1980) 487-494.
92 O. P. Mahajan and P. L. Walker, Jr., Fuel, 58 (1979) 333-337.
93 Y. D. Yeboah, J. P. Longwell, J. B. Howard and W. A. Peters, Ind. Eng. Chem. Process Des. Dev., 19 (1980) 646-653.
94 H. D. Franklin, W. A. Peters and J. B. Howard, Prepr. Div., Fuel Chem., Am. Chem. Soc., 26 No. 2 (1981) 121-132.
95 P. L. Waters, Nature, 178 (1956) 324-326.
96 D. W. van Krevelen, F. J. Huntjens and H. N. M. Dormans, Fuel, 35 (1956) 462-475.
97 V. T. Ciuryla, R. F. Weimer, D. A. Bivans and S. A. Motika, Fuel, 58 (1979) 748-754.
98 D. Fitzgerald and D. W. van Krevelen, Fuel, 38 (1959) 17-37.
99 A. G. Sharkey, Jr., J. L. Shultz and R. A. Friedel, in R. F. Gould (Ed.), Coal Science, Advances in Chemistry Series No. 55, ACS, Washington D. C. 1966, pp. 643-649.
100 A. F. Granger and W. R. Ladner, Fuel, 49 (1970) 17-25.
101 R. L. Bond, W. R. Ladner and G. I. T. McConnell, in R. F. Gould (Ed.), Coal Science, Advances in Chemistry Series No. 55, ACS, Washington D. C., 1966, pp. 650-665.
102 D. M. L. Griffiths and H. A. Standing, *ibid*, pp. 666-676.
103 A. H. Strom and R. T. Eddinger, Chem. Eng. Progress, 67 No. 3 (1971) 75-80.
104 Liquefaction and Chemical Refining of Coal, A Battelle Energy Program Report, Battelle Columbus Laboratories, 1974.
105 Char Oil Energy Development, Office of Coal Research R&D Report No. 73, FMC Corp., 1974.
106 Occidental Research Corporation, Final Report on DOE Contract No. EX-76-C-01-2244, 1979.
107 R. Loison, A. Peytavy, A. F. Boyer and R. Grillot, in H. H. Lowry (Ed.), Chemistry of Coal Utilization, Suppl. Volume, Wiley, New York, 1963, pp. 150-201.
108 L. H. Hamilton, Fuel, 59 (1980) 112-116.
109 L. H. Hamilton, Fuel, 60 (1981) 909-913.
110 T. Matsunaga, Y. Nishiyama, H. S. Wabe and Y. Tamai, Fuel, 57 (1978) 562-564.
111 P. L. Waters, Fuel, 41 (1962) 3-14.
112 N. Y. Kirov and J. N. Stevens, Physical Aspects of Coal Carbonization, University of New South Wales, Sydney, 1967.
113 P. L. Walker, Jr. and O. P. Mahajan, in C. Karr, Jr. (Ed.), Analytical Methods for Coal and Coal Products Vol. I, Academic Press, New York, 1978, pp. 125-162.
114 O. P. Mahajan, in R. A. Meyers (Ed.), Coal Structure, Academic Press, New York, in press.
115 H. Gan, S. P. Nandi and P. L. Walker, Jr., Fuel, 51 (1972) 272-277.
116 P. L. Walker, Jr., L. G. Austin and S. P. Nandi in P. L. Walker, Jr. (Ed.), Chemistry and Physics of Carbon Vol. 2, Marcel Dekker, New York, 1966, pp. 257-371.

162

117   E. C. Harris, Jr. and E. E. Petersen, Fuel, 58 (1979) 599-602.
118   A. Cameron and W. O. Stacy, Austr. J. Appl. Sci., 9 (1958) 283-302.
119   N. Y. Nsakala, R. H. Essenhigh and P. L. Walker, Jr., Fuel, 57 (1978) 605-611.
120   S. P. Nandi, V. Ramadass and P. L. Walker, Jr., Carbon, 2 (1964) 199-209.
121   Y. Toda, Fuel, 52 (1973) 36-40.
122   Y. Toda, Fuel, 52 (1973) 99-104.
123   M. D. Gray, G. M. Kimber and D. E. Granger, Combustion & Flame, 11 (1966) 399-400.
124   D. Anson, F. D. Moles and P. J. Street, Combustion & Flame, 16 (1971) 265-274.
125   D. B. Anthony, J. B. Howard, H. C. Hottel and H. P. Meissner, Fuel, 55 (1976) 121-128.
126   D. B. Anthony, J. B. Howard, H. C. Hottel and H. P. Meissner, Fifteenth Symposium (International) on Combustion, The Combustion Institute, Pittsburgh, 1975, pp. 1303-1317.
127   A. F. Mills, R. K. James and D. Antoniuk, in J. C. Denton and N. Afgan (Ed.), Future Energy Production Systems, Hemisphere Publishing Co., Washington D.C., 1976.
128   R. K. James and A. F. Mills, Letters in Heat and Mass Transfer, 3 (1976) 1-12.
129   P. E. Unger and E. M. Suuberg, Eighteenth Symposium (International) on Combustion, The Combustion Institute, Pittsburgh, 1981, pp. 1203-1211.
130   P. C. Lewellen, M. S. Thesis, Massachusetts Institute of Technology, 1975.
131   D. W. Van Krevelen and J. Schuyer, Coal Science, Elsevier, Amsterdam, 1957.
132   A. Attar, A.I.Ch.E.J., 24 (1978) 106-115.
133   G. R. Gavalas, P. H. Cheong and R. Jain, Ind. Eng. Chem. Fundamentals, 20 (1981) 113-121.
134   G. R. Gavalas, R. Jain and P. H. Cheong, Ind. Eng. Chem. Fundamentals, 20 (1981) 122-132.
135   W. B. Russel, D. A. Saville and M. I. Greene, A.I.Ch.E.J., 25 (1979) 65-80.
136   D. B. Anthony and J. B. Howard, A.I.Ch.E.J., 22 (1976) 625-656.
137   G. J. Pitt, Fuel, 41 (1962) 267-274.
138   G. Borghi, A. F. Sarofim and J. M. Beér, Prepr. A.I.Ch.E. 70th Annual Meeting, New York, 1977.
139   H. A. G. Chermin and D. W. Van Krevelen, Fuel, 36 (1957) 85-104.
140   T. G. Martin and D. F. Williams, Prepr. Royal Society Meeting on New Coal Chemistry, Stoke Orchard, 1980.
141   E. M. Suuberg, W. A. Peters and J. B. Howard, Fuel, 59 (1980) 405-412.
142   R. W. Hiteshue, R. B. Anderson and M. D. Schlesinger, Ind. Eng. Chem., 49 (1957) 2008-2010.
143   R. W. Hiteshue, R. B. Anderson and S. Friedman, Ind. Eng. Chem., 52 (1960) 577-579.
144   M. J. Finn, G. Fynes, W. R. Ladner and J. O. H. Newman, Prepr. Div. Fuel Chem., Am. Chem. Soc., 24 No. 3 (1979) 99-110.
145   R. A. Graff, S. Dobner and A. M. Squires, Fuel, 55 (1976) 109-112.
146   S. Dobner, R. A. Graff and A. M. Squires, Fuel, 55 (1976) 113-115.
147   P. T. Fallon, B. Bhatt and M. Steinberg, Prepr. Div. Fuel Chem., Am. Chem. Soc., 24 (No. 3 (1979) 52-63.
148   J. L. Beeson, D. A. Duncan and R. D. Oberle, Prepr. Div. Fuel Chem., Am. Chem. Soc., 24 No. 3 (1979) 72-81.
149   M. I. Greene, Prepr. Div. Fuel Chem., Am. Chem. Soc., 22 No. 7 (1977) 133-146.
150   C. L. Oberg, A. Y. Falk, G. A. Hood and J. A. Gray, Prepr. Div. Fuel Chem., Am. Chem. Soc., 22 No. 2 (1977) 185-196.
151   P. S. Virk, L. E. Chambers and H. N. Woebocke, in Massey (Ed.), Coal Gasification, Advances in Chemistry Series, No. 131, Am. Chem. Soc., New York, 1974, pp. 237-258.

152 J. M. L. Penninger and H. W. Slotboom, in Albright and Crynes (Eds.), Industrial and Laboratory Pyrolysis, ACS Symp. Series, No. 32, Am. Chem. Soc., New York, 1977, pp. 444-456.
153 G. Schaub, W. A. Peters and J. B. Howard, Proc. Int. Conf. Coal Sci., Düsseldorf, 1981.

AUTHOR INDEX

SUBJECT INDEX